P9-CBB-906

FORCES IN PHYSICS

Titles in Greenwood Guides to Great Ideas in Science
Brian Baigrie, Series Editor

Electricity and Magnetism: A Historical Perspective
Brian Baigrie

Evolution: A Historical Perspective
Bryson Brown

The Chemical Element: A Historical Perspective
Andrew Ede

The Gene: A Historical Perspective
Ted Everson

The Cosmos: A Historical Perspective
Craig G. Fraser

Planetary Motions: A Historical Perspective
Norriss S. Hetherington

Heat and Thermodynamics: A Historical Perspective
Christopher J. T. Lewis

The Quantum Revolution: A Historical Perspective
Kent A. Peacock

Forces in Physics: A Historical Perspective
Steven N. Shore

FORCES IN PHYSICS

A Historical Perspective

Steven N. Shore

Greenwood Guides to Great Ideas in Science
Brian Baigrie, Series Editor

GREENWOOD PRESS
Westport, Connecticut • London

Library of Congress Cataloging-in-Publication Data

Shore, Steven N.
Forces in physics : a historical perspective / Steven N. Shore.
p. cm. – (Greenwood guides to great ideas in science, ISSN 1559–5374)
Includes bibliographical references and index.
ISBN 978–0–313–33303–3 (alk. paper)
1. Force and energy—History. 2. Physics—History. I. Title.
QC72.S46 2008
531′.6—dc22 2008018308

British Library Cataloguing in Publication Data is available.

Library of Congress Catalog Card Number: 2008018308
ISBN: 978–0–313–33303–3
ISSN: 1559–5374

First published in 2008

Greenwood Press, 88 Post Road West, Westport, CT 06881
An imprint of Greenwood Publishing Group, Inc.
www.greenwood.com

Printed in the United States of America

The paper used in this book complies with the
Permanent Paper Standard issued by the National
Information Standards Organization (Z39.48–1984).

10 9 8 7 6 5 4 3 2 1

Dedica:
a **Sephirot**
la forza del mio destino
tu mi hai insegnato a vivere.
Fabrizio De Andre

Contents

List of Illustrations

SERIES FOREWORD

The volumes in this series are devoted to concepts that are fundamental to different branches of the natural sciences—the gene, the quantum, geological cycles, planetary motion, evolution, the cosmos, and forces in nature, to name just a few. Although these volumes focus on the historical development of scientific ideas, the underlying hope of this series is that the reader will gain a deeper understanding of the process and spirit of scientific practice. In particular, in an age in which students and the public have been caught up in debates about controversial scientific ideas, it is hoped that readers of these volumes will better appreciate the provisional character of scientific truths by discovering the manner in which these truths were established.

The history of science as a distinctive field of inquiry can be traced to the early seventeenth century when scientists began to compose histories of their own fields. As early as 1601, the astronomer and mathematician Johannes Kepler composed a rich account of the use of hypotheses in astronomy. During the ensuing three centuries, these histories were increasingly integrated into elementary textbooks, the chief purpose of which was to pinpoint the dates of discoveries as a way of stamping out all too frequent propriety disputes, and to highlight the errors of predecessors and contemporaries. Indeed, historical introductions in scientific textbooks continued to be common well into the twentieth century. Scientists also increasingly wrote histories of their disciplines—separate from those that appeared in textbooks—to explain to a broad popular audience the basic concepts of their science.

The history of science remained under the auspices of scientists until the establishment of the field as a distinct professional activity in the middle of the twentieth century. As academic historians assumed control of history of science writing, they expended enormous energies in the attempt to forge a distinct and autonomous discipline. The result of this struggle to position the history of science as an intellectual endeavor that was valuable in its own right, and not merely in consequence of its ties to science, was that historical studies of the natural sciences were no longer composed with an eye toward educating a wide audience that included nonscientists, but instead were composed with the aim of being consumed by other professional historians of science. And as historical breadth was sacrificed for technical detail, the literature became increasingly daunting in its technical detail. While this scholarly work increased our understanding of the nature of science, the technical demands imposed on the reader had the unfortunate consequence of leaving behind the general reader.

As Series Editor, my ambition for these volumes is that they will combine the best of these two types of writing about the history of science. In step with the general introductions that we associate with historical writing by scientists, the purpose of these volumes is educational—they have been authored with the aim of making these concepts accessible to students—high school, college, and university—and to the general public. However, the scholars who have written these volumes are not only able to impart genuine enthusiasm for the science discussed in the volumes of this series, they can use the research and analytic skills that are the staples of any professional historian and philosopher of science to trace the development of these fundamental concepts. My hope is that a reader of these volumes will share some of the excitement of these scholars—for both science, and its history.

Brian Baigrie
University of Toronto
Series Editor

PREFACE

The history of physics is the conceptual history of force. It is also the story of how quantitative analysis became ever more central to, and effective for, exploring physical phenomena. Thus, a few basic questions will flow as a theme throughout this book: "What is motion?", "Why is there motion?", and "How is there motion?" These are so elementary that they may strike you as trivial. After all, you know how to describe the motion of a ball rolling along a bowling alley or flying through a stadium. You know that it is moving (it's changing its location relative to you and its target), why it is moving (it was thrown or struck or kicked by someone, probably intentionally), and how it is moving (it has some speed and goes in some direction). But woven through the tapestry of the argument in this whole book will be an effort to make such commonplace observations terribly difficult because those who struggled with them for so many centuries thought they were and that's the only way to step into that now vanished world. In these amazingly ordinary questions you find the origin of physics, the effort to understand the laws governing the structure of the material world that spans more than two millennia from the time of the Greeks to the present.

The order of presentation is historical, beginning in antiquity, but don't mistake chronology for linearity. The difficulty in discussing the conceptual development of science is how it proceeds simultaneously along many lines of inquiry. Sometimes random coincidences strike a resonant chord and produce new, unexpected, results. Philosophers of science have puzzled over this for generations. These may come from the accumulation of paradoxes, problems, and contradictions during a period between great leaps, or they may arise from simple "what if" questions. The history of force displays both modes of discovery. But science in general, and physics in particular, is the product of communication among those who engage in this grand activity. It is a collective activity with a rich history. If we think there are really laws of nature then, in one sense, it doesn't actually matter who finds them—eventually they'll be discovered. For a working physicist, attempting to understand and wring something fundamental out of experiments and observations, theory is about how to best represent these laws and the phenomena they produce. It is about models, how we approximate reality. That's easily stated but I remember reading this sort of statement when I began studying science and it made no sense at all. Now, I think, it's because the presentation was often completely ahistorical. The laws seemed to come out of nowhere. The same thing usually happened in mathematics. Worse still was the presentation of the "scientific method," which seemed almost silly. Why would you ever think of testing something, just as an idea out of the blue? How are the hypotheses formed in the first place and why? I hope you'll find some explanation of that in this book.

I should explain what I won't be doing here. There are several other books in this series that deal with the development of astronomical, cosmological, chemical, and biological ideas. You should approach this work keeping in mind its broader setting. While I discuss some of the developments in these areas, the presentation is intended to be contextualizing rather than complete. Some parts will seem sketchy, especially when describing planetary theory from the time of the Greeks through the later Middle Ages. That's intentional and would seem to require a separate book, as indeed it does in this series. It's difficult to write a purely descriptive history of any physical science, but this is particularly true for mechanics. It was the font of much of modern mathematics, and the symbiosis that marks its development forms an important part of the story. A word to the reader is in order regarding the subsequent increasingly mathematicized discussion of this book. At the end of the seventeenth century, a new method was introduced into physics, the infinitesimal calculus or, as it is now called, *differential* and *integral* calculus. The mathematical advance was to recognize that the limit of finite impulses, for instance, could produce in an instant of time a finite acceleration or displacement and that it is possible to describe a continuous variation in any quantity. For motion and force this can be regarded as the most important step in the entire history of physics. Equipped with this new tool, Newton's law can be written in a continuous, invariant form that redefines acceleration and renders impulse a limit of a continuously acting force. One-dimensional histories can be expanded to three, and changes explicitly relative to time and space can be separated. Further, the history of a motion—a trajectory—can be described with the continuous sum called an integral. This is just an example, there are others and I've added a short appendix that gives a few more details. I'll make no apologies for the increasing use of equations and symbols here. In fact, those who are seeing this for the first time will, I hope, find an inspiration for further study in this mathematical physics. Finally, a word about the biographical material and choices of personalities. Many thinkers contributed to the development of physical ideas and it's not my intention to make this a set of capsule biographies set within a simple timeline. Instead, while I will focus on a relatively few individuals, the ideas are the principal aim of the work and I hope the reader will not find the irregular path traced by the text to be too confusing. The names and a few key concepts provide the landmarks, but not the milestones, in our journey. What for me has always been the richness of the study of the history of science is how it reveals the recurrence of the fundamental problems, often in unexpected and beautiful ways.

ACKNOWLEDGMENTS

I've benefitted from advice and criticisms from many colleagues while writing this book and it is a pleasure now to publicly repay that debt. I want first to thank Brian Baigrie, the editor of this series, for his kind invitation to participate in the project and for constant encouragement and precious critiques. Jason Aufdenberg, Pier Giorgio Prada Moroni, and Laura Tacheny provided detailed comments and gentle nudges throughout the process. Sincere thanks also to Claude Bertout, Federica Ciamponi, Craig Fraser, Tamara Frontczak, Daniele Galli, Stefano Gattei, Enore Guadagnini, Pierre Henri, Michael Kinyon, Ted LaRosa, Ilan Levine, Monika Lynker, Patrice Poinsotte, Veronica Prada Moroni, Paolo Rossi, Gabriele Torelli, Roberto Vergara Caffarelli, Lesley Walker, and Lyle Zynda for advice and many valued discussions about the topics in this book. Antonella Gasperini, at INAF-Arcetri, Firenze, and Gabriella Benedetti, Dipartimento di Fisica, Pisa, were always ready and willing to provide bibliographic help. I also thank Oliver Darrigol and Tilman Sauer for correspondence. Kevin Dowling, as editor for Greenwood Press, displayed the kindness and patience of a saint and the wisdom of a sage as this process has extended past every deadline, and I also thank Sweety Singh, project manager at Aptara Inc., for her valuable help during the publishing process. I'm grateful to Edwin and Anne DeWindt for their friendship, counsel, and inspiration as historians and scholars.

And thank you Cody.

This book is dedicated to the memory of two great scholars and teachers with whom I had the privilege of studying during my graduate studies at the University of Toronto, Fr. James Weisheipl and Prof. John Abrams. They lived their subject and through their love of learning transported their students into a timeless world. They instilled a critical spirit through their questions, discourses, and glosses, always emphasizing how to see the world through the eyes of the past. The long road of science is a tangle of errors, commentaries, corrections, and insights, of misapprehensions and clarifications, and most of all a communal adventure that I think is well summarized in a passage from *Adam Bede* by George Eliot:

> Nature has her language, and she is not unveracious; but we don't know all the intricacies of her syntax just yet, and in a hasty reading we may happen to extract the very opposite of her real meaning.

1

FORCE IN THE ANCIENT AND CLASSICAL WORLD

> But Nature flies from the infinite, for the infinite is unending or imperfect, and Nature ever seeks an end.
>
> —Aristotle, *Generation of Animals*

Although the civilizations of Mesopotamia and Egypt certainly knew and used elementary force concepts, at least as they were needed for what we would now call engineering, they appear to never have systematized this experience within a theoretical framework comparable to their developments of geometry and astronomy. For this reason I'll begin with the Greeks. They were the first to attempt to form a unified explanatory physics—a science of Nature—that combined quantitative scaling methods with a set of more or less mundane principles, taken as axioms, that produced a universal tradition of extraordinary persistence and elaboration. It's precisely this "ordinariness," you could say "common senseness," of the rules that appeared so persuasive, as we'll see, and that also made Greek physics so incredibly difficult to supersede. Indeed, it can be argued that several centuries' pedagogy has not yet led to its eradication.

For the ancient authors, the problem of motion *was* the problem of force. But there is a sort of historical determinism in this statement: for the whole subsequent history the problems of mechanics have been intimately connected with how to describe time, space, and motion. Since we'll begin with classical and Medieval arguments, it is easy to think these are quaint and almost empty disputes and conjectures from a distant past about things that are obvious to anyone. One of the points of this book is to show they are *neither* obvious nor distant.

It seems so obvious: you push, pull, or twist something and you've applied a force. It reacts: it moves, it deforms, it swerves. But why does the thing you're "forcing" react? How does it react? Under what conditions will it react? And if it doesn't, why? Add to this the problem of motion—what it means to say something is "changing" or "displacing"—and you have the catalog of the fundamental questions. There was also a theological motivation and it had to do with the state of the world as it was created. Recall the statement, for instance, in Genesis, "it was good." That which is "good" is perfect and that which is perfect is unchanged. If the world created by the gods were in harmony then it should be in balance, in equilibrium. The problem is the world isn't static, we see motion and change. Things form, creatures are born and die, the seasons pass and the Sun moves across the sky. Either all these changes are illusory or they have a cause. To make matters more difficult, the rate of change isn't the same among all things. Changes of location are different than changes of state. Things move in space at different speeds, and in various directions, and age—that is, move in time—at different rates. Living things are automotive, acting or changing because of innate abilities and desires. Inanimate bodies also move, but here the cause is more obscure. It seems to require some external agent, but it isn't obvious that the reaction of the body to this agent's actions will always be the same. If you think about this for a moment, you'll see why the problem of motion—and therefore of force—was so difficult to resolve for so long. You need to recognize that Nature isn't always animate and, then, that it isn't capricious or malevolent. You also have to find some way to distinguish between "voluntary" and "obligatory" actions and then you can begin asking how and why these occur.

How can we describe change? It seems to require memory, that from one instant to the next we recall that something had a previous state (location, appearance, etc.) and now different. The cognitive act, recalling a previous state and then seeing that the thing in question has moved, is central to our notion of causality and is a hard-learned lesson in childhood. The infant learns to recognize what is an essential property of a body, what makes something the same as it was regardless of how its location changes. This notion of the stability of things and will become progressively more important. To say something changes spontaneously, without apparent reason or *cause*, is to say it acts inscrutably. This is often applied to people, who have *ideas* as agents for their actions, and in a supernatural world it can also apply to things. To see that, instead, there are systematic behaviors—"laws" of nature—was an enormous step.

Let's start with a simple example. Imagine a stone lying on the ground. Now it just stays there unless *you* physically do something to it. Telling it to jump into your hand doesn't work, except perhaps in animated cartoons. Its place must be actively changed. So let's say, you lift it. Now you notice it feels heavy. You can move your hand in the same way without the stone and lift the air, but the difference in weight is obvious. Now you release the stone. It falls. Obviously there's nothing unusual about that, right? But think again. To get the stone to rise, *you* had to do something. What are *you* doing when the stone falls? Nothing, that's the problem! If the stone isn't alive, how does it "know" where to go. An even greater puzzle comes from throwing it upward. Again you are the active agent but the stone slows and then reverses its direction once it leaves your hand. Why does it continue moving and what makes it "change its mind"? To complicate the picture, at the next level we can ask why some objects fall—the rock on the edge of a table, for instance, if it's disturbed—while others never do—fire, for example. Some things fall faster than others, so it seems how motion occurs is somehow linked to the nature of matter. For all ancient peoples, the gods were always available to explain the big events, from geopolitics to natural catastrophes, but for such ordinary things of so little importance it seemed ultimately ridiculous to require their constant attention to each attribute of the material world. Instead, things themselves can be imbued with motives and potentials. You can substitute here the construct of a "world soul" and get closer to the idea: spirits are linked somehow through a sense of purpose to produce the required state of the world according to the master plan. The Platonic notion of perfect forms, and of the Demiurge, the purposeful creator who sets the whole world out with a reason and then steps back knowing how it will develop *any why*, is enough to explain anything without contradictions. If something seems extraordinary, for better or worse, it can be put down to the parts of the design that are unknowable by humans and therefore simply there without requiring deeper probing. This deistic worldview has no problem with any cause because it is inherent to the thing in question having been carefully provided by the design at the start. The remaining problem of first cause, the *origin of the Designer*, is left as an exercise for the reader, although we will soon see that this too produced problems and resolutions in the original mechanical systems.

Starting from a purposeful world, motion of a body is, in itself, not mysterious. For instance, if a living creature wants to move, it moves. There is some internal animation, caused by the "mind" or "will" which, when coupled to the "machine" of the body, produces a change in the being's state of rest or motion. If the world were animated, if there were a "world soul" as the agent for change in the natural world, then motion would always intentional and, one might say, reasonable. That's all that's required in a supernatural context. But it becomes strange when willful cause is replaced by some action of the world on the thing in question without "motivation." We call this new concept *Force* and the boundary between spontaneous and willful motion and obligatory response to the "laws of nature" marks our starting point in this journey. Force as a causative agent to motion is the core concept of classical physics. No other study illuminates so clearly the confrontation of experience with the physical world.

THE SETTING OF ANCIENT PHYSICS

The Greeks were the first civilization to develop the notion that Nature is comprehensible through logic and based on only a limited number of principles. Starting with Thales, around 700 BC, they were drawn to philosophical—that is, non-theistic—explanations of phenomena in the world around them. Before the fifth century BC, Anaxagoras and Anaximander were performing demonstrations from which they drew physical conclusions, the beginnings of empiricism, and Pythagoras had applied mathematical reasoning to the discovery that musical harmonies are related to number. Among the philosophers before Plato, the Pre-Socratics, we find a wide range of speculation about causes. None of these are more than suggestions, they were frequently couched in metaphor and analogy. Even Plato, in the fourth century BC, dealt more with perception, logic, and mathematical reasoning—the discovery of abstraction as a way of understanding how to penetrate beneath appearances to the nature of things—than with physical causes.

At Athens, Plato attracted remarkable talents to the Academy. His circle included the geometers Euclid and Eudoxus. And then there was Aristotle, who will be the focus of our discussion of ancient Western physical thought. There are several reasons for this. He was the first to create a systematic body of works extending across the whole spectrum of physical experience, the first to develop a comprehensive method of analysis—the dialectic—and the most complete source we have for his predecessors. But it is perhaps more important that *later* writers began with Aristotle. For the Scholastics, he was "the Philosopher," the font of their methodology and the foil of their arguments. He was the source for the major problems of physics, among many others, and his is the largest surviving collection of treatises on physical subjects from the Hellenistic age.

NATURE AND MOTION: THE *PHYSICS* AND *DE CAELO* OF ARISTOTLE

A characteristic feature of ancient natural philosophy that, persisted into the Renaissance, was a preoccupation with classification—taxonomy—by which a thing can be described and consequently known. This approach, which works remarkably well for zoology and botany, was one of the inspirations for Aristotle's approach. You first classify the thing being studied. For example, you ask "what kind of motion are we seeing," the idea that there are different "species" of a physical phenomenon is at the base of the method. Then following the enumeration of the detailed properties of the subject, through a series of "compare and contrast" exercises, you arrive at its essential properties.

Two works, *Physics* ("On Nature") and *de Caelo* ("On the Heavens"), present Aristotle's ideas about earthly and celestial mechanics. Above all, for him motion *is* the problem of force, and vice versa, and thus it remained for nearly two thousand years. Aristotle's way of explaining motion can be viewed as a transformation of choice and motivation into physical necessity. His mechanical ideas were set within the broader context of taxonomy: to find the agents by defining the actions. Thus the basic distinctions among the causes is between those that are intrinsic to the body and those that relate to its state. These causes, the fundamental elements of Aristotle's explanations, are enumerated very early in the *Physics*, in the second book:

- *Material*: anything that exists must be composed of something material, so the stuff out of which a body is produced contains something causative.
- *Formal*: the thing also has some form, some shape, some components, and these are causative as well. Formed differently, the material will act differently.
- *Efficient or effective*: something or someone creates the thing, so there is a causation in this. Here we meet for the first time intention in the creative act, and at the end of the sequence we have
- *Final*: the ultimate, purposeful end toward which the thing acts. In other words, once formed, there is a final state that the thing should be in.

As a naturalist, Aristotle took his analogies from the living world. It is, therefore, not surprising that within the four causes, the *material* has the potential within itself to produce any action if it is properly "stimulated." This latency becomes a property of a substance. How this metamorphosed into a universal, far more general idea of "potential" we will see once we get to the eighteenth century. But in its ancient context, this "Potency" is very different from a force, or mass. It is a cause within the thing, something proper to the nature of the object that can even include its form as well as its substance. While the action isn't obviously voluntary, it almost resembles a magical property, an animate nature, that if understood can be controlled and that stays with the individual thing. There is nothing universal about the things we see in the world except their composition: if you take the basic elements, that have specific properties in themselves, and combine them to form the material world, the things thus formed will retain mixtures of those basic "natures."

In order, then, to understand how things move and change it's necessary to also have some idea of what matter is and how these properties mix within something. For Aristotle, there were four fundamental elements: earth, air, fire, and water. Each possesses specific, distinct properties that imparted certain predictable behaviors to matter. Substance is continuous, there are no fundamental indivisible particles, *atoms*, from which everything is constructed, and the essential ingredient is the mix of these elements. Time is a "magnitude divisible into magnitudes," as he affirms in Book 3. Any interval can be infinitely subdivided; since the present is an instant, there must be an infinite past and an infinite future. The same holds for space and, by combination of space and time, all motion is continuous, so only a finite motion can occur in a finite time. Like magnitude, or line in geometry, motion is infinitely divisible. In its division, there are two distinct types of motion, "violent" or "accidental" and "natural" that are essentially different. All matter is either simple or compound in nature depending on the mix of the four elements, each is dominated by the appropriate natural motion of the principal components. If you imagine the analogy with living things, this is very intuitive. The taxonomy is based on recognizing the proportional distinctions between creatures and plants. But this also applies to inanimate things. For example, you can imagine a body that is water and earth—that is, mud. If dropped, it will follow the path of the dominant element, earth, and seek it's appropriate place, the center. A cloud can shoot fire, rain, and rise so it too must be a compound of fire, water, and air.

The reasoning is actually remarkably successful in distinguishing among the things of the material world. It even seems predictive of their behaviors. Each of the elements has its natural motion and its necessary destination of a natural place. These might, however, be the same. For instance, although air is transparent and light, it has weight (this was already well known for at least a century before Aristotle, but fire doesn't. Yet smoke is a mixed thing that comes from releasing that element from a body. The heaviness and its natural place, down, permitted—even required—one more concept within Aristotle's physics, the most sweeping conclusion of his whole system. Because all heavy things—that is, all things endowed with *gravitas* (weight)—fall to the ground, and because from anywhere on Earth you see the same action, since the Earth is a sphere it must be at the center of the world. From the principle of "natural" motions, it also follows that the Earth must be immobile because it must seek its own center; a substance cannot have within itself more than one natural motion so rotation or orbital motion are both ruled out. Thus we have a complete world picture from the outset, a physical basis for the constructions of the astronomers and mathematicians, a cosmology.

Instead, "violent motion" is impressed on a body by an agent that causes it to act against its natural motion or that disrupts it from its natural state. Thus, any time you cause something to move contrary to its expected behavior the motion must diminish and ultimately end. While natural motion only terminates when the body reaches its appropriate place, violent motion stops of its own accord when the action of the moving agent stops. A willful, even random or *accidental* cause can produce any motion you'd want, but the agent can't make something act against it's nature forever and its influence must be expended. The natural motion will eventually dominate.

TIME, PLACE, AND CHANGE

How do you separate the space in which a body moves from the body itself? For Aristotle, this was subtle. One needs to imagine something like an "external" space, indeed space itself is a very difficult thing to realize if you cannot imagine a vacuum. To have an empty space that is *not* merely an abstraction, something that isn't just a vacuum but genuinely nothing but a coordinate reference frame, requires being able to put a physical thing—like a material body—within a nothingness. Then, to have this body move, we can ask if the space that contains something is separate from the thing it contains? And if something changes its state, if it moves or varies in one or more properties, how is the change to be described? Even more puzzling is how we know that something *has* changed.

All change becomes relative and takes place in time. One of Aristotle's predecessors, Zeno of Elea, found this puzzling. Since something can be called different by simple juxtaposition with another state, irrespective of when the two states occur, we could imagine motion occurring in the same way. These are his (in)famous paradoxes of motion. Taking just one of his examples, the flight of an arrow, we can ignore the cause of the motion and simply ask if it is possible for the arrow to be moving. Of course, you'll say, we have two positions and just compare where the arrow is relative to them. But for Zeno, instead of taking this external view, he poses the problem as one seen from within the moving frame. The famous, and in its time thorny, problem of "Achilles and the Tortoise" is posed this way. If Achilles starts from point A and the Tortoise at B, then when the Tortoise moves from B to C, Achilles moves from A to B. Similarly with all displacements. Since the two points bounding the displacement are never coincident, the conclusion is the Tortoise is safe; Achilles can never capture it. But Zeno has cleverly left out the *time* required to move through the respective intervals. Having space alone isn't enough, motion requires time: if we ask how fast Achilles moves relative to the Tortoise we come to a very different (perhaps unfortunate) conclusion for the hapless reptile and agreement with our senses. Thus in the third book of *Physica*, Aristotle makes time the central argument, along with the nature of actions, reaction, contact forces, and the infinite while in Book 4, he treats place, void, compulsory *versus* voluntary motions, and again the nature of time.

Space and time were whole in themselves, but for Aristotle there were "types" of change. Motion isn't only spatial displacement. It could also be an alteration of its current condition, a change in something's *state*. If the world were, for instance, in equilibrium then nothing would vary. But we see change of many kinds so we have to ask why this happens. You can see here the argument against Parmenides and his school who, in the generations before Aristotle, had affirmed that all change is illusory and purely a product of the senses. Instead, we now have a distinction of three types of change. First, as the term implies, *accidental* changes are those events that just happen, although not necessarily by chance. They can be intentional acts or haphazard events; it suffices that they are not something produced by the thing in motion on its own, nor by its essential properties. This contrasts with *essential* change that depends on the nature of the matter, the particular mix of characters. The last is *self-motion*. While it's almost obvious for living things, it's the hardest for Aristotle to explain for "stuff" without resorting to some animating principle. To flag a point now that we will soon face in more detail, when recovered in the Latin West, this Aristotelian idea became the link between theology and cosmology. The self-moved thing, the *Primum mobile* or first mover, is perpetual and perfect, and therefore was identified later with an animate God.

Ironically, it was just this identification that Aristotle and his followers, the Peripatetics, sought to avoid. Although there are natural motions, there is a difference between matter in general and living things. The first are moved by principles—laws—while the second are moved by desires and intentions. Stuff, matter, is merely obedient to laws and not capricious. The Prime Mover is also bound in action by these laws and does not act in an intelligent way to control the system, it just does what comes naturally. Being composed of a distinct substance than terrestrial things, it cannot be found in the stuff of the everyday world but it is constrained to behave according to the laws that govern everything.

THE THING MOVING

Now, finally, Aristotle can treat the motion of the affected body and the nature of corporeal motion, whether the part or the whole is moved, and the impossibility of moving an infinite distance in a finite time. Again, these are set within a specific picture of how the force is transmitted to the object which is always through some contact force.

In book 7, Aristotle makes the statement that would echo down the centuries as *the* axiom of motion. He states that *everything that moves is moved by another. For if it does not have within itself the power of moving, it is evidently moved by another*. Although this seems circular at first, it is an essential separation between things that have a natural motion or are animated and those that experience violent or accidental motions because something outside has started it and maintains it. Freely falling bodies do not have an obvious driving agent yet they move—this was the point about something having within itself the power of motion. For example, imagine a rock sitting on a shelf. If the shelf breaks, the impediment to the natural motion is also removed so the motion starts. It continues to fall until the ground gets in the way and stops the motion. The rock comes to rest because it isn't able to overcome the resistance of the surface. Both of these depend on circumstance, so they're accidental in Aristotle's view. But the rock, being heavy, contains within itself the natural cause: it seeks the center of heaviness, the center of the world. In contrast, an animal, is born (comes into being) with its willful ability to move as part of its nature. No external driving is required.

How the mover acts and what it ultimately is were questions that Aristotle could not avoid and in the *Physics* and in *de Caelo*, it is at the core of cosmology toward which the whole of the *Physics* has been directed.[1] Thus, in the last section of the ultimate (eighth) book of *Physics*, Aristotle at last tackles the problem of first motions and their initiation. If all motions require something else to initiate them, there must be a first agent or "Prime Mover." The theological and philosophical implications of his solution dominated physical theory for almost two thousand years.

CELESTIAL MOTIONS

The problem of planetary motion was distinct. Unlike the stuff of everyday life, the planets—and this includes the Sun and Moon—revolve unceasingly. The stars never change their motions, staying fixed relative to each other and revolving from east to west daily. They neither fall nor slow down. The planets, on the other hand, are more complicated. They too follow the daily motion of the stars but displace relative to them.[2] Most of the time they move in the same direction as the Sun, opposite the rotation of the stars, from west to east along the same path followed by the Sun. But occasionally, periodically, they reverse this motion and pass through a short period of *retrograde* during which they stop, reverse their angular motion for a period of weeks to months, stop again, and then continue as before. The angular motion is not constant even in the prograde part of the revolution. For the Moon and Sun, the variation is substantial, for the others (except Mercury, for which the motions are especially complicated) it is barely observable and required a long period of observations and tables. The irregularities of the lunar and solar motions were already well known to the earliest astronomers, the Babylonians and Egyptians, had codified the kinematics in a set of laws from which they could construct tables. The Babylonians had developed a system for computing the solar and lunar wanderings among the fixed stars, treating the rates of motion as a function of time through simple algorithms that changed the angular speed depending on the position relative to the Sun. But notably, neither they nor the Egyptians hypothesized any physical explanation for these motions. If the tables worked well enough to cast horoscopes and determine time, that was enough. As to their essential natures, celestial bodies remained inscrutable.

The Greeks were the first to combine these individual cases into a single explanatory physical system. Eudoxus, a contemporary of Plato and Euclid in the fourth century BC, is credited with the first kinematic explanation for the motions using geometrical methods. First, he assumed all motions are centered on the Earth that sits unmoved at the center of the sphere of the fixed stars. This sphere rotates once per day around a fixed axis. Then he asserted that the mobiles are carried on nested, variously tilted geocentric spheres; the number varied depending on what was required to *reproduce* the motion of each planet. The *model*, and here we encounter—really for the first time—an attempt to create a general explanation of all possible motions of the celestial bodies, required alternating directions of rotation so the combined effect could not only produce the normal west-to-east motion of the planets relative to the stars but also explain the daily motion of the sky as a whole *and* the occasional reversals of the planetary progressions.

These intervals of *retrograde* travel cannot be accounted for without compound motions.

Aristotle adopted this system essentially unaltered but sought to add a driving mechanism. Not content to merely mimic the motions, he sought a cause and found it in the contact of rigid spheres with a complicated analog of a transmission system of a machine. Sets of compensating spheres, increasing in number as new phenomena were included, could reproduce the basic kinematics while remaining consistent with the dynamics. Most important, since the "infinite" is inadmissible in Hellenistic cosmology, there must be something maintaining the driving at a finite distance from the center. This was the *primum mobile* or Prime Mover. For Aristotle this was the outer sphere was the origin of the motions and had within it the formal and material causes for driving the motion of those spheres within it.

STATICS: ARCHIMEDES

If the science of motion was the stuff of philosophy within mechanics, statics was its applied side. The development of technology had always been practical and by trial and error, with artisans learning the hard way how to produce, and then reproduce, mechanical effects. Machines were inherited and modified conservatively,

Figure 1.1: A Roman mosiac from the first century AD, discovered at Pompei, depicting the death of Archimedes during the Roman siege and conquest of Syracuse. Image copyright History of Science Collections, University of Oklahoma Libraries.

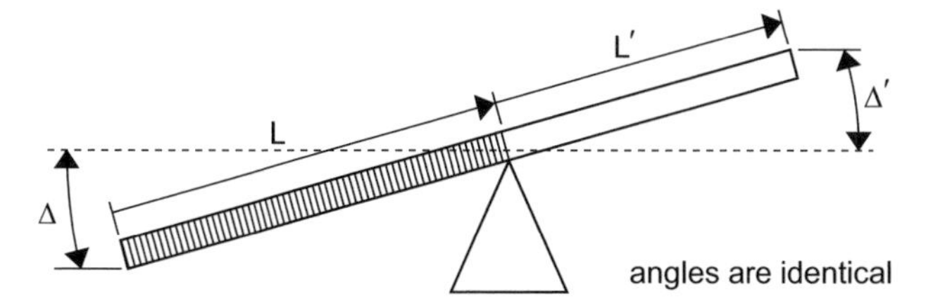

Figure 1.2: The level. Displacement of the left downward moves more mass at a larger distance from the fulcrum (the support point) through the same angle than the right side (the vertical distance on the left is greater than that on the right).

almost like natural selection. Once a function was fixed, they were varied only hesitantly. The builders were known by their skills, which were acquired by experience. But the Greek philosophers were a questioning bunch and even technical matters excited their curiosity. The systematics of these rules of thumb were like laws of nature. Mechanical contrivances were, after all, operating according to mechanical principles and it became a challenge to find out what those were. Aristotle didn't say much about the equilibrium of continuous media in the *Physics*. The point of the work was to examine motion (although in book 4 of *de Caelo*, he dealt briefly with floating bodies). A separate tradition began with the *Mechanical Problems*, in which equilibrium and change were treated together. This work, which is no longer considered Aristotle's own but that of one of his students (probably Theodorus, who followed him as head of the Lyceum) dates from the period immediately after Aristotle, at the end of the fourth century BC. Its schematic presentation resembles lecture notes expanding on basic issues from the *Physics* and relatively brief responses to a wide range of mechanical questions. The work was unknown in the Latin West but it was quite popular among Islamic scholars and consequently indirectly exerted a broad influence on doctrines that would later jump the Bosporus into Western arguments.

The principal developer of these ideas in the ancient world was Archimedes, who was born in Syracuse in 287 BC and died during the Roman siege of the city in 212–211 BC. His was a unique combination of talents in mathematics and applied mechanics, what we would now think of as the prerequisites for being a successful engineer. But I'll call him the first "physicist" in a sense very different than the other Greek philosophers of nature. He had the peculiar ability to extract broad theoretical principles from mundane problems and to then turn these around to produce devices and practical applications. His particular concern was not with motion but statics. In effect, he was the first to see rest, not just motion, as an important subject of inquiry and the first to apply quantitative, mathematical reasoning to bodies in equilibrium.

Aristotle had asserted that once a body is in its natural place it is at rest and, therefore, in equilibrium. Instead, for Archimedes, equilibrium isn't something that just happens. It's when opposing forces, whatever that means, balance. So it's not surprising that the balance, or more precisely the *lever*, provided the necessary example for Archimedes. Our main source for his thoughts, as it was for the Islamic and Medieval philosophers, is his *On the Equilibrium of Planes or the Centers of Gravity of Planes*. His approach was deductive, although his axioms were those of an applied natural philosopher and geometer. His presentation is worth examining at some length. He begins with a set of axioms the principal one of which is "Equal weights at equal distances are in equilibrium and if not equal, or if at unequal distances, the greater weight if unequal or the one at the greatest distance if equal will displace the plane downward." Then he presents essentially the same assumption but with a new condition, what happens when a new weight is added to the system so it goes out of balance and then uses more or less the same but now if a weight is subtracted. This is typical also of the later presentations in the Middle Ages, each case is stated separately even if they amount to the same thing with changes in the signs (for instance, a weight subtracted is a negative weight added, a difficult concept before algebra). He then assumes that "Equal and similar figures coincide applied to each other as do their centers of gravity" and follows the same procedure asserting that "In figures unequal but similar the centers of gravity are similarly placed. Here, however, an additional construction is added to the axiom, that if points are similarly situated and connected by similar lines they will be in equilibrium. A fundamental new element is added when he proposes to quantify the arguments by introducing the word "magnitude" to indicate the weight of a body. Finally, the location of the center of gravity must be within the figure if its sides are concave in the same direction. This last axiom, again purely geometric, provided a very important notion in his studies. With this assumption Archimedes could specify how to distinguish between "interior" and "exterior" components of a body.

Archimedes then extends these axioms in the propositions using the style of geometrical reasoning, gradually building the general result by examining special cases. For instance, the third proposition, *unequal weights will balance at unequal distances if the distances from the fulcrum are inversely proportional*, is the basic law of the lever. Prop. 4 combines figures to determine their common center of gravity if the figures are self similar or joined by a line to their respective centers of gravity. Prop. 5 extends this to three bodies along the line connecting the centers of gravity. Prop. 6 and 7 restate, more precisely, prop. 3 now not using commensurate figures (in other words, here he uses the notion of dead weight). Prop. 8 is the same as the level for the interior of a figure, thus beginning the definition of equilibrium of a solid within itself. prop. 9 deals with the equilibrium of parallelograms, Prop. 10 makes this precise to the statement that the center of gravity is at the center of the figure (intersection of the diagonals of the parallelogram). Prop. 11 deals with similar triangular planes, extended to the location of the center of gravity more precisely in prop. 12 and 13. These are summarized with prop. 14, that *the center of gravity of any triangle is at the intersection of the lines drawn from any two angles to the midpoint of the opposite sides, respectively*. This was later extended in the *Method*. The final prop. 15, is the culmination of the triangular figure but makes use of the parallelogram construction. The second book opens with the consideration of parabolic segments and then parallels the presentation of the first book. In effect, having established the construction for simple figures, the point of the second book is to demonstrate its extension to arbitrary, curvilinear figures of which the parabola happens to be an analytically tractable case. Thus, using purely geometric constructions, Archimedes was able to generalize the treatment of balance to a broader class of bodies.

The importance of his concept of equilibrium is more evident in his next work, as far as they can be sequenced chronologically, *On Bodies Floating in Water*. The moment of discovery is the stuff of legend, told by Vitruvius, the first century Roman mechanician—Archimedes noticing the rise and fall of the level of water in a bath as he lowered or raise himself and in a fit of excitement, running naked through the streets of Syracuse yelling "eureka" (I've found it). Whatever the truth of the story, the observation was correct and profound and sufficient to solve the problem of how to determine the density of an object more precisely than could be affected with a simple beam balance. In his theoretical treatment, Archimedes discusses the application of a balanced state of rigid bodies immersed in a deformable incompressible substance, liquid water. The treatment is a beautiful application of the lever, taking the center of mass of a figure, in this case a paraboloid (a surrogate for a boat), and treating the *limiting cases of stability*. For the first time, we are presented with *constraints* within which a body can maintain its balance. This is *not*, however, a stability argument. Archimedes doesn't discuss how a body reacts, or how it regains its equilibrium, or whether it continues to move away from that state, nor what happens to its motion as it does, but only under what conditions the body will reestablish its balance. In all cases, there is no discussion of the motion resulting from thrusts and drives. The new principle is relative weight, the free air weight of a body as measured by a balance reduced by the weight of the water displaced when it is immersed.

A material object in balance under its own weight is described in terms only of gravity and moment. Weights or magnitudes, which are the equivalent in the Archimedian literature, generalizes the results to extended bodies. The center of gravity becomes not the center of figure, which may take any shape, but the center around which the figure is in equilibrium and that consequently produces its entire effect as a weight. While this seems simple, it was a distinct change from the classification of formal causes that dominated much of the rest of physics. The body can have its weight redistributed in any manner along an equilibrium line as long as it doesn't move its center of gravity. Archimedes uses this in the *Method* and also in the two mechanical works on equilibrium.

This is the part of the mechanical tradition that intersects experience, especially regarding the construction and stability of structures. Archimedes applied it to a wide range of common problems, from the balancing of beams to the equilibrium of floating bodies. In this second, he stands as the founder of naval architecture. Not only are we dealing with the ability of a body to stay afloat, that follows from an application of the equilibrium principle of the level, but its ability to stay upright dependent on shape and loading can now be addressed. Here the lever is paramount: the location of the center of gravity and the length of the moment arm determine whether any body can remain stable.

Not only machines were now the object of physical investigation. Even construction was encompassed in the Archimedian approach to materials. A beam is just a lever and an arch is a distributed load along a curved beam. The monuments constructed by the Egyptians, Greeks, Romans, Chinese, and Indian builders have survived millennia, a testament to the impressive body of empirical knowledge accumulated by ancient and Medieval architects. No theoretical treatises exist on any of these structures. The Hellenistic compilation by Hero of Alexandria, dealing with mechanical inventions along the lines of Archimedes, and the later Roman work of Vitruvius and Medieval Vuillard di Harencourt describe a wide variety of machines and designs but generally with no need to invoke calculations. Based on the simplest geometric forms, since the builders were unschooled in geometry and could deal easily only with right angles, circles, and polygonal inscribed or circumscribed figures, nonetheless yielded a rich heritage of structures and architectural forms. This lack of theoretical basis for the building process was no barrier to the realization of the projects. It meant a skilled class of artisans developed over the centuries and that the accumulated experience formed a tradition rather than a formalized body of knowledge. Nonetheless, this provided a background for many of the later discussions.[3]

Applied Hellenistic Mechanics: The Basic Machines

When Archimedes wrote about thrust and driving, he was *not* using the terms in their modern sense (although it sometimes seems so from translations). Instead, the thrust is the weight of the body and the reaction is that of the fluid. Because the medium, water, is incompressible there isn't any need to discuss the structure of matter or any effects of the actions on the bodies themselves. In this sense, he could deal *geometrically* with equilibria. But lacking any driver, or it seems any interest in what happens *outside* of balanced forces, he was content to deal only with those configurations that are stable. For instance, in the treatment of the righting of a floating paraboloid, he solved a basic problem of naval architecture—where to locate the center of mass of a vessel and under what conditions it would be stable (depending on the load), but what happens if things go out of kilter is not Archimedes' problem.

That is more a problem for those we would now call engineers, the artisans and technicians who actually made real working things. The Hellenistic mechanicians of the first century, especially Hero of Alexandria, and their Roman successors, especially Vitruvius, began their discussions of mechanisms in much the same fashion as Aristotle divided the causes in his discussion of motion, by distinguishing five from which virtually all others derived. These were the *block and tackle* or *pully*, the *wedge*, the *screw*, the *winch*, and most important of all, the *lever*. For all, the same principle applies, mechanical advantage is the efficiency with which a smaller force can overcome a larger one. The lever provided the main example of equilibrium of extended bodies, the pulley showed how a force can be distributed. Again, all that was needed was weight (loads) and reaction (tension) without any clear idea of forces. For instance, a cord provides the tension to lift a body and, in general, this was treated without extension. A single wheel suspended from a rigid beam or ceiling allows you to redirect the force but changes nothing about its magnitude—you still need to exert W units of force for each W units of weight. But if a second pulley is added and has a weight attached to it, then the cord extending over the two with one end attached to a rigid support reduces by a factor of two the amount of force required to lift W. Adding more wheels to the system further reduces the required force, each provides a factor of two reduction. This was described by Hero in his *Mechanics* and became the block and tackle. For the other mechanical elements the treatments are similar.

THE BIRTH OF THEORETICAL ASTRONOMY

Hellenistic science achieved a complete picture of natural phenomena in the terrestrial realm within Aristotle's physical system. But the motions of the Moon, Sun, planets, and the stars presented different but related difficulties to those in the nearer parts of the world. Let's begin with the sky. The stars move continually in the same direction and, apparently, with the same angular motion. But nothing else does. This perpetual motion is completely alien to the physics we have just outlined. Aristotle and those who followed saw at least one immediate solution to this dilemma. If the universe is governed by the same causes everywhere, they might differ in their effects depending on site. So it was straightforward to add another postulate to the system. Here around us, we see things changing. Weather provides a perfect example. Clouds form and dissipate, it rains, there are winds from various directions at different times. While these must originate from the same principles as the abstractions of single particle motion and equilibrium, it is also obvious that the celestial bodies don't show this same lack of predictability. There is nothing of chance in the heavens. This can be included in a general scheme by separating the universe into two parts, those within the atmosphere and those without.

Although the Moon changes shape, this was obviously due to its relation to the Sun and merely an effect of reflection of light. Otherwise it behaved like the Sun and planets, moving perpetually around the Earth and relative to the stars. Because the Moon eclipses the Sun and occults stars and planets, it obviously lies closer than they do. Its distance was known from direct parallax measurements, at least in Earth radii, before the start of the first century and as early as beginning of the third century BC Archimedes and Aristarchus of Samos had proposed lower limits for the distance to the Sun. The sphericity of the Earth and its central location were supported by astronomical observations, and even the relative size of the lunar to the solar orbit could be estimated from the duration of lunar eclipses. The fixed stars, however, were a different issue. They couldn't be infinitely far away; Aristotle had argued, sufficiently convincingly, that an infinitely fast motion cannot occur in nature and since the heavens revolve in a finite time they cannot be infinite in extent. But unlike the Moon, and the Sun, no simple geometric measurement or argument could furnish a stellar distance scale. It was therefore reasonable to proceed by taking the Eudoxian spheres as the fundamental mechanism, for planetary motion and make the outermost sphere that transports the stars and planets in their daily motion the outermost of the universe.

Celestial Motions

The Classical universe was divided into two parts: "down here" and "up there," the things of everyday life and the things in the heavens. Nothing connected them directly, leaving aside the notion of "influences" that served as the wrapping for astrological prognostication.[4] The distinction between the two realms was almost tautological. The celestial was eternal, the terrestrial wasn't. If one of these heavenly things fell, for instance meteors, it was reclassified. Any motion that seemed to share properties between the two regions, comets for example, were thought to be *essentially* something terrestrial in origin in which, somehow, the properties were mixed. Freefall and celestial circulation were treated differently but both were, in Aristotle's sense, natural since they were "proper and essential" to the bodies that display these motions.

Circular motion required a special treatment. While it was straightforward to postulate a teleology for the fall of a body, or some similar explanation for the loss of impulse in violent motion with increasing distance from the source, circular motion and its continuation posed more complex problems. It required instead formal and material causes to sustain the motion, the perfection of the circle (or sphere) and the nature of the substance from which the celestial realm is formed. The separation of the heavens from the Earth made it plausible that the matter would also be distinctive and therefore able to do things you wouldn't expect to see happening in the world below. This separation, which is an essential feature of all pre-modern physics, was reasonable in the absence of general principles for the initiation of motion and the transmission of action.

Freefall provides the empirical basis for cosmology and also the context for explaining the continued circular motions of the planets and other extraterrestrial bodies: the universe is centered on the Earth, which stands immobile. This was as true for the mechanicians as it was for the astronomers. Archimedes, for instance, took this for granted and used it to explain the equilibrium of layers of fluid.

You should distinguish the dynamical and kinematic explanations. For describing the structure of the world, cosmology needs a consistent principle. The motion is the only thing that matters: you don't need to ask dynamical questions because all motions are identical, simply circular. Thus the kinematics reduce to just geometry, albeit with the addition of time as a reference. The relative motions are important for astronomical observation and the chosen reference is to the Sun.

A few months' observation, if properly timed around a planetary opposition when the planet is in the part of the sky opposite to the Sun (meaning we restrict attention to Mars, Jupiter, and Saturn in order of their periods) is enough to see that these bodies exhibit deviations from a steady progression along the ecliptic. These epochs of *retrograde* motion, again being purely kinematic, can be reduced to geometric analysis as Apollonius (third century BC) was the first to clearly demonstrate. The kinematical proof makes no use of the concept of acceleration. All apparently irregularities must be due to compounded circular periodic motions. How these add is a separate matter from their origin. The motions are *continuous*, it's only our particular vantage point that makes them seem to be accelerating. With this axiom, and the cosmological assumption of terrestrial immobility, it was possible to construct a model of the universe and a single surviving work by the Alexandrian geometer Claudius Ptolemy (flourished mid-second century) shows how it was accomplished: the *Mathematical Composition* (in Greek, *Syntaxis*). Written in the middle of the second century, it became the most important source for theoretical astronomy among the Islamic mathematicians, for whom it was the "The Great Book," by which name it became known in the Latin West when it was translated in the twelfth century by John of Seville: *Almagest*. But it isn't with the mathematical constructions that we'll be concerned. This is also a work also provides a record of how Aristotelian physics merged with cosmology to create a world system for the cosmos.

Ptolemy, because he was writing a work on the motions of the planets, needed to discuss the construction of the heavens. His physical premises are stated in the introduction to *Almagest*, asserting that immobility of the Earth underlies all other principles of celestial *mechanics*. If the planets are to continue in movement, they require constant driving: they move because they are moved. The mechanism is contact, the agent is spherical, and the driver is external and perpetual. For Aristotle, and for all subsequent commentators the ultimate cause for the motion was a property, or principle, of the Prime Mover. In this system, to find a cause for the motions requires ending what would otherwise be an infinite regress. The first cause must be something that has this action as part of *itself*. The ancient astronomers and geometers, particularly Apollonius, Hipparcos, and Ptolemy, all faced a problem. The mean motions are regular. But you see deviations that are systematic and periodic in both space and time. These behaviors could be represented by successively approximating the motions, by adding motion-specific mechanisms to the orbit in a way that produced ever more refined periodic time series for the kinematics.

Let's take a moment to examine how this construction worked. By assertion, any motion must be either regular or compounded from regular motions. This holds for the periods as for the directions and rates of motion. Since a body can't act "contrarily" to its nature or capriciously, the reversal of directions and changes in the rates of motion of the planets must be produced by a combination of regular, sometimes counteracting, motions. The first motion, the *mean anomaly*, is the time for an orbit relative to the fixed stars. This was the principal circle or *deferent*. If the motion varied in a regular way, speeding up and slowing down periodically around the orbit, it sufficed to displace the observer from the center of the circle, the *eccentric*. If further variation was observed, other points of relative motion could be added, the first was the *equant*, on the opposite side of the center from the eccentric. Since the Sun moves only in one direction with variable speed throughout the year this worked quite well: the *observed* angular velocity varies but the *actual* angular motion remains constant. But there are problems with this picture: the planetary motions are not identical. Although the Sun and Moon move only from west to east in ecliptic longitude, albeit with variable speed, they never reverse as do three of the planets: Mars, Jupiter, and Saturn. Also, relative to the Sun, Mercury and Venus also reverse direction. That is, they stay locked to the solar motion, never deviating by more than some fixed angular distance from the Sun despite the variation in the solar angular speed. This is the basis of epicyclic motion. The *epicycle* is a circle of fixed radius and inclination that carries the body around a separate, moving point on the *deferent*.

The stationary points and reversals in direction of motion of a planetary trajectory is due to the relative motions adding in such a way that the body appears to be at rest, a proportion depending only on the ratio of the relative angular motions on an epicycle and that on the deferent and not on any real change. This was demonstrated by Apollonius of Perga in the third century BC using a purely geometric construction. But the success within Ptolemy's model comes with a price. This kinematical construction is inconsistent with the dynamical principles used to justify its necessity: the epicycle requires motion about at least one separately moving point that is not simply connected to the driving spheres and, consequently, to the Prime Mover. This violation of cosmological construction led to a separation of physical and mathematical astronomy very early in its development and was responsible for many paradoxes, some of which were later resolved by the Copernican simplification. For Ptolemy, as for other astronomers, the axioms were physical but these could be violated by constructions *if necessary*. The debate has been heated about the meaning of *model* and *hypothesis* in ancient astronomy but I think this much is clear. If the planets do or don't move along epicycles was not really a question that concerned the ancients. On the one hand, all treatises including and following the *Almagest* begin with a section outlining cosmology. This is a philosophical prelude. It sets the stage for the problem to be treated, but being mathematics—which treats ideal things—the methods used cannot *explain* the motions. For that, only metaphysics (and later theology) will do. Ptolemy says as much in the *Almagest* and later, in the *Tetrabiblos*, repeats the assertion that the models of the heavens are aimed at describing and predicting the motions within some set of assumptions or rules:

> But it is proper to try and fit as far as possible the simpler hypotheses to the movements of the heavens; and if this does not succeed, then any hypotheses possible. Once all the appearances are saved by the consequences of the hypotheses, why should it seem strange that such complications can come about in the movements of the heavenly things? (Almagest, Book 13, part 2)

If these contradict, the physical picture tells us that the description is incomplete and we have yet to find the most efficient and consistent model. We need not give up on building a picture because it is just the shadow of the "real," what we can grasp as a calculation is how something appears and not what it "is." In this I think the requirement of the models to *preserve the appearances* shouldn't be seen with modern hindsight. Our notion of a model has evolved through the last 500 years, separated in some ways irreversibly from the ancients. What the Renaissance left was a reinterpretation of Platonism and a new concept of the use of mathematical reasoning in physical problems that in many ways doesn't relate to those of the ancients.

In particular, noting that the first mover must have a separate status in this system (as Aristotle made clear in Book 8 of *Physics* and Book 4 of *de Caelo*), the cause of the motion is always present. The problem was that the system was that each motion required a mechanism, and these were added according to a prescription that requires replicating the phenomena rather than remaining concordant with the *Physica*. The lunar orbit is a perfect case in point. The combination of eccentricity—the variation in the angular speed of the Moon without having retrograde motions—and the systematic shift of the phase of the maximum speed forced the addition not only of an eccentric location for the center of motion but a second, moving point, the equant, around which the speed was constant. Neither was associated with the center of the spheres that were presumably driving this motion. The same was used for Mercury. Even more seriously discrepant was the intersection of the epicycles of Mars and Venus. There was no scale in this system so the angular motions, and distances from the Sun, require proper relative proportions of the deferent and the epicycle. Angular size variations, which should have been obviously wrong to even casual observers for the Moon, were simply ignored. So were the relative angular sizes of the Sun and Moon during eclipses. The restriction that a celestial mechanics be physically as well as phenomenologically self-consistent wasn't part of the thinking.

THE STRUCTURE OF THINGS: ATOMISM AND MOTION

There was another view of the world besides the Aristotelian continuum, atomism, that provided alternative, radically different explanations of natural phenomena. The problems of the structure of matter and the origin and maintenance of motion came together for the atomists. This line of thought, descending from Democritus (end of fifth to beginning of fourth centuries BC) and Epicurus (around 342 BC to 270 BC, a late contemporary of Aristotle), proposed that material and void coexist and are everything. Matter is atomic, transient, and mutable. It's formed from the temporary cohesion of atoms, which are complicated in their interactions and various in composition, and when this fails to be stable, the material thing decays and returns to the dispersed medium. There is nothing other than atoms so the universe isn't filled by any resisting medium.

Beginning with Democritus, but having even earlier origins, the idea that matter is based on an almost unimaginable number of different and specific, indivisible elementary units: atoms. Our only complete exposition of this view of matter, and of the forces that maintains it, comes from the lyrical composition *de Rerum Naturum* ("On the Nature of Things") by the Roman poet Titus Lucretius Carus, a follower of Epicurus, in the first century BC. Although we know next to nothing precise about him, his poem remains a rich source for the ideas that, when re-introduced in the Latin West, provoked a debate on the foundations of the world that echoes to our time. How atomism contrasted with the prevailing continuum picture is most dramatically exhibited in the nature of space and forces. For Aristotle and his school, the world is filled with matter, the vacuum is impossible, and motion takes place only because a force overcomes its resistance. For the atomists, in contrast, the universe is empty except for the myriad interacting individuals that, in their unceasing motion, randomly unite and separate. Motion requires no explanation in itself, matter is, instead, its principal result. All interactions depend on the nature of the atoms themselves, whose properties are specific to the qualities they bring to the things formed. Those of the "soul" are different than those of, say, a tree. How they interact is left unspecified, although it seems the force is thought of as impulsive during collisions. It is enough to say the atoms are eternal and the things of the world are transient. That which we see is produced by chance. It's tempting to see in this ancient idea as something similar to our worldview. We're now so accustomed to a discrete world at the microscopic level that the atomist doctrine that you may feel more at home with it than that of Aristotle and the filled world. But as we will see much later, this picture is *analogous* to the modern conception of the microworld only in the broadest sense. There is no concept here of what a force is, the motion has an origin that begins "somehow" and there is nothing here to explain how the individual atoms differentiate to correspond to their material properties.

The Islamic Preservation

Hellenistic science passed into the Islamic world rather than to the West during the long period of political and social instability that characterized the dissolution of the Roman empire at the end of the fifth century and the establishment of stable kingdoms and communes after the Ottonians in Germany and the Carolingians before the start of the millennium. Although a few works survived and were preserved in a few copies, it is clear that the intellectual fervent of the Islamic courts and centers of learning were debating and extending the ideas and techniques of this inheritance, especially in Spain. The philosophers were interested in calculational and practical problems, much less in theoretical mechanics, and their advances in formal analysis and mathematical technique was enormous, but that's a separate development from our main line. In dynamics, they generally followed the Aristotelian precepts. Islamic mechanics used, and made significant advances on, Archimedian results and methods. In naval design, in particular, but also in discussions of machines and balances the principles of hydrostatic equilibrium, specific gravity, center of gravity, and the lever as a paradigmatic physical device, all found their expression in the texts. In astronomy, they elaborated the Ptolemaic system. Thabit ibn Quura (836–901) modified the treatment of the motion of the sphere of the fixed stars (the eighth sphere) contained in the *Almagest* by adding an cyclic motion to explain what was thought to be a variable rate for the precession of the equinoxes. Ibn Sina, known in the Latin West as Avicenna, wrote about atomism and forces. For ballistic motion, Ibn Rashud, later known as Averroes, invoked the action of the air as a means for propelling bodies in their flight, a kind of hydraulic forcing that maintained motion after a violent action. They discussed many of the cosmological problems in theological and philosophical terms but made no essential changes to the explanations. A prime example is the extended discussion of astronomy in Maimonedes' work *The Guide for the Perplexed*, one of the clearest available expositions in the Islamic tradition of Aristotle's principles.

NOTES

1. This question of how the motion is transmitted is the most contested issue in the long literature on the history of Aristotelian physics. Nothing in any of Aristotle's works, however, details *how* the action is imparted to the body.

2. The Sun and Moon are distinct since they are the only resolved objects that execute the same periodic motion as the planets.

3. Our discussion has focused on the Hellenistic discoveries. But these were not unique. A striking parallel comes from the third century in China, as reported in the *San Kuo Chih*:

> The son of Tshao Tshao, Chhung, was in his youth, clever and observant. When only five or six years old, his understanding was that of a grown man. Once Sun Chhüan had an elephant, and Tshao Tshao wanted to know its weight. He asked all his courtiers and officials, but no one could work the thing out. Chhung, however, said, "Put the elephant on a large boat and mark the water level; then the weigh a number of heavy things and put them in the boat in their turn (until it sinks to the same level)—compare the two and you will have it." Tshao Tshao was very pleased and ordered this to be done forthwith.

There were two separate weighing operations for alloys—one in air and the other in water—that were used as a test for purity as early as the Han dynasty (from the second century BC to the third century). Also during the Han, brine testing was performed using floating bodies, for instance eggs and lotus seeds. The latter are particularly interesting because a much later description dating from the twelfth century records that the shape of the seed was also used to determine the relative salinity of the water. If the seed floated upright, the brine was a strong solution, while for horizontal equilibrium the brine was weaker. These and other examples discussed in the compendium on Chinese science and engineering by Joseph Needham (1962) show that the Chinese experience with hydrostatics was considerable although without the theoretical basis that characterized the Archimedian methodology.

4. I can't pass this by completely. The astrological tradition flourished throughout the ancient world and frequently was invoked to justify astronomical investigations, for instance that a better understanding of planetary motions would improve the accuracy of predictions, and the practical product of mathematical astronomy was a rich compilation over several thousand years of almanacs for such purposes. There was no contradiction in this activity, and no separation, precisely because there was no general physical law that prevented such correlations.

2

MEDIEVAL IDEAS OF FORCE

> Each of the four elements occupies a certain position of its own assigned to it by nature: it is not found in another place, so long as no other but its own natural force acts upon it; it is a dead body; it has no life, no perception, no spontaneous motion, and remains at rest in its natural place. When moved from its place by some external force, it returns towards its natural place as soon as that force ceases to operate. For the elements have the property of moving back to their place in a straight line, but they have no properties which would cause them to remain where they are, or to move otherwise than in a straight line.
>
> —Book 72, *Guide for the Perplexed*, by Moses Maimonides, Friedländer tr. [1904]

For the Greeks and Romans, physics has meant the analysis and explanation of natural phenomena. It sometimes found supernatural applications, for instance alchemy and astrology where it seemed capable of providing precise predictions, whatever their actual success. But more often it just satisfied the philosopher's curiosity and explained links between phenomena. With the introduction of these ideas into Islamic culture and then their transmission to the Latin West during the Middle Ages, this changed and so did the questions. In Europe, in particular, the audience that received Aristotle's works and the later commentaries was concerned with very different questions than those that had motivated their author, in fact, they were diametrically opposite. Instead of seeing principles as things in themselves, the clerics who first read and elaborated on the works of Greek science had ideological goals in mind, seeking and finding in them a justification of a religious doctrine. For many centuries that followed the turn of the millennium, to discuss physics—especially mechanics and astronomy—meant discussing God.

I need to emphasize this point. Some of the problems posed in the medieval texts seem almost naive. They often mix kinematics, dynamics, and metaphysics. For example, whether the Eucharist can be ever present and yet consumed, whether

God can be in all places at once, whether action is direct or at a distance, how things change and yet remain constant, were all theological as well as physical questions. The transformation of these questions first into grammar and logic and then into geometry took hundreds of years of debates and small modifications. The appearance of a new school of thought in the fourteenth century, one that began to create a mathematicized picture of motion and force, should be seen as an outgrowth of these more speculative questions.

THE SETTING OF PHYSICAL SCIENCE IN THE MIDDLE AGES

In the twelfth century the Latin world discovered, or rather recovered, the intellectual production of the Hellenistic scientists and the result was explosive. For centuries, only fragments had been known of these ideas, preserved in those Latin commentaries that had managed to survive almost randomly from the end of the Roman empire. Few of these were at a level above the most elementary, many were corrupted redactions, and none were sufficiently detailed to permit any extension. The period we now call the Dark Ages was hardly dark, but it was largely illiterate and there were much more serious concerns of daily life to contend with than would permit the leisure necessary to consider the abstractions of mathematics and physical science. It's important to consider this when confronting the later developments. Ours is a world so interconnected by written and visual communication it is likely very difficult to imagine how isolated the intellectual production was during the period before the introduction of movable type in the West. Even in the Islamic world, more privileged in its possession of many of the Greek products that had been translated into Arabic early enough to promote a flourishing of commentators and innovators, individual works were circulated in manuscript, single copies made from single texts. Their diffusion was slow and limited to certain environments that had access to the production and discussion. A single work could take months to produce, depending on the contents. The few great ancient libraries, especially that in Alexandria, had been destroyed or dispersed, and those into whose hands the surviving texts had fallen often could do nothing with them. Some, perhaps the best example of which is the *Method* of Archimedes, had been reused as writing material or even binding for other folios.[1] In the mid-thirteenth century, William of Moerbeke translated a huge body of Hellenistic works into Latin, especially those of Plato, Aristotle, Archimedes, and Euclid, working directly from the original Greek although not the original manuscripts themselves. This made works previously unavailable accessible to scholars. Other scholars used Arabic or Hebrew translations of the Greek works, still others translated original Islamic works into Latin. Wider circulation of these texts took some time, but within a century they had found their way into collections and libraries throughout Europe.[2]

Another feature of the medieval texts is notable. They are frequently commentaries, not extensions. A scientific paper is now placed in its developmental context by a brief introduction, often just a series of references to indicate the development of the idea, with the main part of the paper—or book—being new developments. Novelty is now the distinguishing feature, not brilliance of exposition of ideas already known. The concept of received wisdom was very important in the Middle Ages and the texts were often approached as relics to be treated with respect rather than criticism and correction. Every manuscript was not merely an artifact, it was almost sacred. The time required to produce a single work could be very long and it is clear that many of the copiers, and compilers, were unaware of the subtleties of the material with which they were dealing. The climate was not one prone to rapid innovation and the long, slow, redundant production before the rise of the Scholastics and Commentators and the beginnings of the universities in Europe was in sharp contrast to the active Islamic production that was building on the Greek texts. This is especially important following the two condemnations of teaching of Averroist and Avicennan doctrines against either Aristotle or scripture, both issued in Paris in the thirteenth century. The first, in 1210 AD, forbade the teaching of certain doctrines as true. The second, in 1277 AD, was issued at an ecclesiastical level low enough to be chilling on the Parisian scholars but not damning all academics and not extending throughout Europe, Like its doctrinal predecessor, it only prohibited taking a list of propositions as "true," not discussing and disputing them (as long as the outcome of the debate was predetermined). That's why I mentioned the structure of the commentary. This insured the propositions and many, if not all, of the arguments were presented to facilitate their refutation. Thus we know considerably more about the texts because the edict was simply a ban, not demanding the destruction of the works (condemned texts were burned, a perfect way to prevent the spread of the precise text of a heretical work in an era when texts may be represented by only a single manuscript).

One last point before we proceed. A reason for the centuries-long silence was not ignorance but disinterest allied with an attempt to rigidly enforce centralized thought and a deeply conservative hierarchy. With the rise of the universities in Europe in the thirteenth century things changed dramatically. These were no longer isolated clerical centers of textual production and discourse. Instead they were international centers, collecting students and teachers from all over Europe. You can see this same polyglot environment now in any of their descendents all over the world but in the two centuries from 1200 to 1400 they were still novel places, loosely organized among themselves and often competitive for students and faculty. In the same way as the cathedrals, they were seen as prizes of the communities in which they were sited, although not without considerable "town and gown" tensions. Their history is beyond the scope of this book but their place in the growth of science in general, and physics in particular, cannot be overestimated. The faculty lectures were not merely delivered but also written, copied, and diffused. The scribal houses, which in their time resembled the clusters of photocopy shops that now surround every college and university, served as depositories for the intellectual production of the faculty and students. They received licenses from the university to "publish" the lectures deposited there and the universities themselves preserved the notes of the courses in their libraries, which were more accessible than those of the monasteries. Writing materials also changed, passing from parchment to paper, and the decreasing costs or reproduction and growth of the literate population eager for new texts and receptive to new ideas spurred change amid a new intellectual climate.

Of course, science was not the principal subject of study, but it was an essential element of the curriculum. Mathematics, specifically geometry, was seen as fundamental to train the mind in precise reasoning along with logic. Philosophy was not only speculative or theological but also concerned with nature. Astronomy (and astrology), medicine, anatomy and botany (and alchemy), and physics all formed subjects of study and debate. For this reason we find a significant shift in the centers of activity and the affiliation of scientific authors with schools. Beginning with Salerno, Bologna, Paris, Oxford, and Cambridge, between 1200 and 1500 many of these centers of learning were founded and flourished, surviving wars and plagues, economic collapses and civil unrest, changing the values, discourse, ideas, and structure of Western society. Not least important was the effect on the sciences, to which we'll now return.

MEDIEVAL STATICS

The problem of balance was as puzzling to the Medievals as it had been to the Greeks, especially the problem of bodies balanced in different angles or on bent levers. This had been treated by Islamic philosophers but it was at the heart of the treatises by Jordanus de Nemore (flourished around 1220 AD) whose works were the basis of the medieval discussion of weight and equilibrium. These works introduced a new physical element, a nascent concept of *work*. It's easy to imagine this as a precursor of the eighteenth century notion we will discuss soon, but although the form of the argument is similar the physics is very different. Jordanus argued this way. Beginning with the lever as a constrained motion, since the attached weights must remain attached and the arm remains rigid, to be in equilibrium for two different weights W_1 and W_2 the distances at which they are placed from the pivot, L_1 and L_2, must have the inverse ratio to the weights. That is, $L_1/L_2 = W_2/W_1$. If displaced from the horizontal, the vertical distance (the length of the arc) will vary directly as their distances. But because these are inversely proportional to the weights, even if the lever is inclined the bodies will remain stationary. Another way of saying this is that the weights must have the proportional distances from the pivot that produces equal work when displaced from the horizontal in either sense.

Another very new concept, *positional gravity*, allowed Jordanus to solve the problem of two coupled weights on an inclined plane, one of which is suspended vertically and the other placed along the plane. He distinguished between this and the *natural gravity* that the body has along the vertical that he showed is always the larger value. The proof of equilibrium is very revealing. He first showed that the *positional* gravity is independent of location along a constantly inclined plane. This he shows using a new, and dramatic, idea that comes very close to the later concept of *work*. But remember, we are dealing here with statics. The displacements are not taking place in time, although the word "velocity" is sometimes used. Instead, this is the first instance of the method of *virtual displacement*. If the distance to which a body must be raised, which is directly proportional to its distance from the pivot in balance, is proportionally greater than the ratio of the respective distances, the weight cannot be lifted.

For a moment, let's anticipate a development that at the time of Jordanus lay several hundred years in the future. Although sometimes thought to be a dynamical proof of the lever, Jordanus was using a very special setup to describe the equilibrium of weights. They're linked, either by a cord or a lever, and thus constrained to move in a very specific way. This is *not* the same as motion on an inclined plane, although it shares many of the same features. Positional gravity is indeed a component of the force, although it was always represented as a virtual vertical displacement, but it is not a force in the sense that Galileo would later understand. As I noted when discussing Archimedes, how a body displaced from its equilibrium returns or diverges from that state is not the issue. It's enough to say that the system is unbalanced. It is for this reason, and this alone, that Jordanus was able to resolve so many problems using only the lever as a paradigm. Constrained motion with inextensible connections requires equal and opposite displacements of the components. If one rests on an inclined plane, it is lifted through the same magnitude of distance but by the projected vertical distance, with respect to another body displaced along some other inclined plane. Because the positional gravity is independent of location along the plane, the initial condition can be to place the two weights at anywhere as long as the length of the cord remains constant.

The idea of work used here is pretty vague but it serves the purpose of quantifying experience: the *relative* effort required to raise a weight is greater for a larger distance, or for a heavier weight raised though any fixed distance. The word *relative* here is very important. In the Archimedian hydrostatic balance, it isn't the absolute weight (in air) that is determined but the relative change for two materials of the same free weight but different density—that is, different specific gravity—that recognizes that the fundamental feature of the balance is that all measurements are made *relative to a standard*. Since all measurements are proportional, with a chosen standard (as would have been done in a market using marked weights) the problem remains without any need to use specific units. For instance, immersion of two bodies of equal weight—in air—produces different displacements in water and the relative densities are therefore simple to determine. In effect, in all these cases, the lever provides the tool for understanding the general problem. Even a pulley can be reduced to a treatment of mechanical advantage, the relative lever arms being the different radii.

Understanding the difference between this static conception of equilibrium and Galileo's later dynamical approach can be illustrated by an experience you've had in childhood. If you recall sitting on a swing and pumping with your legs to get started, the initial oscillation you experienced is what Jordanus was thinking of as positional gravity. You're changing the distance of a weight, your legs, along the axis of the pendulum, the swing has what we would now say is a variable moment of inertia, and you're shifting your center of gravity. This is even more obvious when you lean into the direction of motion. But what happens on a swing is something never discussed in the medieval works. If you hit the right frequency the amplitude of the swing continues to increase; you reach a resonance, but this wasn't realized until nearly 500 years later. The problem of balance is not connected with dynamics even when discussed using times. For constrained motion, as long as the time interval is fixed, displacement and velocity are the same thing and it was only a convenience to use the latter.

The addition of components of force would become central to Galileo's argument for the action of gravity. Using this notion he also was able to discuss the effect of a balanced pair of weights with one oscillating as a pendulum without dealing with the cause of the motion or its properties. It was enough that the change of the position of the swinging body relative to the center of mass changed its effect on the balance, that the angle—the position—changes the weight *as felt in the coupled system*. Thus Jordanus and his commentators could solve the perplexing problem of a bent lever by showing that only the horizontal arm, the *static moment* matters when considering the balance of the weights. The proof given in the treatises is also easily reproduced experimentally and Jordanus even comments on the pitfalls of the demonstration.

The Scholastic treatises are notable for this new approach to how to specify when two bodies are in equilibrium. But it would be a very serious mistake to think that Jordanus and his contemporaries had a more sophisticated concept of force than weight. They were discussing, for instance, displacement, velocity, and force equivocally. This works when you are dealing with a constrained system, such as a machine, but not for free motion. Notably, Jordanus didn't discuss that. It wasn't until Galileo that the two come together. But by introducing components of weight and a static notion of virtual work, the thirteenth and fourteenth century treatises on statics, printed by Peter Apian in the 1530s, provided a fertile source for the development of mechanics in the subsequent centuries.

In contrast to the surviving works of Roman authors, who generally dealt with statics and what we would now call "applied" physics, the medieval analysis of motion seems much more quantitative. It still, however, maintained the same distinctions between kinematics and dynamics we found in the earlier treatments. The concept of a force is missing, although the word is used along with resistance. It has more the nature of *cause*, that is something that produces and maintains the motion, and comes in a variety of types in the same way the motion does. Force is the continued action of a mover, either by contact or somehow, and it produces a motion precisely as long as it is applied.

SCHOLASTIC DYNAMICS: IMPETUS AND ACCELERATED MOTION

Motion was a central issue during the thirteenth and fourteenth centuries, mainly among two schools: at Merton College, Oxford in England, and Paris, in France. These centers of intellectual activity and international contact insured a wide audience, both the actual attendants in the lectures and public disputes held in the universities, and through the circulation of manuscripts emanating from the *scriptoria* connected with the universities. Further, becoming more readily available during the early era of printed book production, especially because of the role of the university presses in the dissemination of texts, these works and their ideas were even more widely distributed through Europe at the end of the fifteenth century.[3]

Acceleration had a special fascination because it involves an alteration in the rate of change in the state of the body. In a sense, all motion is, from the scholastic viewpoint, forced but not all is accelerated. The particular problem was, again, ballistics. An upwardly moving body doesn't continue this motion. After some time (distance), it reverses and falls, continuing to move in this opposite direction without needing any assistance. We find a solution in the works of Jean Buridan. Born around 1300 AD, he lectured in Paris during 1340s and served for a time as the rector of the university; his death was sometime before 1360. He was one of the most influential scholastics and the founder of the Paris school of mechanics. Let's examine, for instance, some examples from Buridan's *Questions on the Eight Books of the Physics of Aristotle*, in this case Book 8, question 12. Here Buridan argues against the idea that motion has its source in the medium, the *antiperistasus* principle.

> 2. . . . The third experience is this: a ship drawn swiftly in the river even against the flow of the river, after the drawing has ceased, cannot be stopped quickly but continues to move for a long time. And yet a sailor on the deck does not feel any air pushing him from behind. He feels only the air in front resisting. Again, suppose that this same ship were loaded with grain or wood and a man were situated to the rear of the cargo. Then if the air were of such an impetus it could push the ship along more strongly, the man would be pressed more violently between the cargo and the air following it. Experience shows this is false. Or at least if the ship were loaded with grain or straw, the air following and pushing would fold the stalks that were in the rear. This is all false.
>
> 5. For if anyone seeks why I project a stone farther than a feather, and iron or lead fitted to my hand farther than just as much wood, I answer that the cause of this is that the reception of all forms and natural dispositions is in the matter and by reason of the matter. Hence, by the additional amount of matter, the body can receive more of that impetus and more intensity. Now in a dense and heavy body, all other things being equal, there is more of prime matter than in a rare and light one. Hence a dense and heavy body receives more of that impetus and more intensity, just as iron can receive more calidity than wood or water of the same quantity. Moreover, a feather receives such an impetus so weakly that it [the impetus] is immediately destroyed by the resisting air. And so also if a light wood or heavy iron of the same volume and shape are moved equally fast by the projector, the iron will be moved farther because there is impressed in it a more intense impetus which is not so quickly corrupted as the lesser impetus would be corrupted. This is also the reason why it is more difficult to bring to rest a large smith's mill which is moving swiftly than a small one.
>
> 6. From this theory also appears the cause of why the natural motion of a heavy body downward is continually accelerated. For from the beginning only the gravity was moving it. Therefore it moved more slowly, but in moving it impressed in the heavy body an impetus. This now acting together with its gravity moves it. Therefore the motion becomes faster; and by the amount it is faster, so the impetus is more intense. Therefore the movement evidently becomes continually faster (translation is from Clagett 1959).

In the next three questions, Buridan outlines the nature of the impetus. First, it is the motion and moves the projectile, but it is caused by the projector because a thing cannot produce itself. Second, that impetus is not simply additive. It is not the local motion alone but a quality of the motion. And finally, "the impetus is a thing of permanent nature that is distinct from the local motion in which the projectile is moved." And "it is also probable that just as that quality is a quality naturally present and predisposed for moving a body in which it is impressed just as it is said that a quality impressed in iron by a magnet moves the iron to the magnet. And it is also probable that just as that quality (impetus) is impressed in the moving body along with the motion by the motor; so with the motion it is remitted, corrupted, or impeded by resistance or contrary inclination." In the last part he uses the example of a stretched cord that is struck. He notes that it doesn't return to its initial position "but on account of the impetus it crosses beyond the normal straight position in the contrary direction and then again returns. It does this many times."

I should make a remark here about the methods employed by the Oxford and Parisian scholars when discussing physical problems. They proposed and discussed several distinct force laws. But none of these were ever compared with experience although at the same time there was increasing attention being paid to the role of empiricism—that is, experience—in deciding philosophical questions. Even when the propositions are apparently quantitative, it is misleading. The numbers inserted in the examples are simply chosen out of thin air, it's actually doubtful that many, or even any, actual experiments were performed by those who wrote the texts; these were arithmetic exercises to illustrate the preconceived laws rather than measurements. For example, William of Ockham (c.1280–c.1349), who is far more famous for his logical treatises than those dealing with physics, nonetheless discussed aspects of the Aristotelian cosmology and mechanics, especially questions about space and time. Roger Bacon (1219–c.1292) and Robert Grosseteste (c.1163–1253) also advocated experiment, mainly for optical phenomena, but the implications for all three was that some laws of nature could be *discovered* through experiment and not merely demonstrated.

Merton College, Oxford was the center of dynamical theory in England. A remarkable group of scholars collected there in the fourteenth century, the most important of whom were Thomas Bradwardine (c.1390–1399), William of Ockham, Duns Scotus (c.1266–1308), Richard Swinsehead (active c.1340–1355), Williams Heytesbury (active c.1335), and John Dumbleton (died c.1349). We'll focus on one aspect of their natural philosophy, the attempt to render the laws of motion in a mathematical form. In this effort they treated not only those explicitly from Aristotle but other attempts to connect motion with force from the previous centuries. Their school, collectively known as the "calculators" because of their quantitative approach, was to be the outstanding influence in the subsequent development of mechanics. Bradwardine was the founder of the research program and the author of its principal text, although he left Merton College in 1330s. His chief work, the *Tractatus de Proportionum*, was written in 1328. Bradwardine's methods extend beyond the simple kinematics of the Mertonians. He suggests several laws of motion, none of which are precisely justified but each of which has special characteristics. The first is the straightforward interpretation of Aristotle's doctrine of resistance, the velocity is proportional to the ratio of the impressed force to the resistance of the medium (whatever that might be). The second is a compound law, that the effect of an impressed force is magnified by the body, always in ratio to the resistance, expressed algebraically as $V \sim (F/R)^n$.

We take as the fundamental dynamical principle that there must be something continuously acting to maintain a state of motion. No motion is natural, in the sense of the spheres, if it is in terrestrial stuff. So as a first assumption, the velocity is a "measure" of the force. But does that mean it is equal to the power of the mover or only an effect? This isn't just a semantic question: it is the core of the confusion among the commentators in the thirteenth and fourteenth century. There were two principles: the force must overcome a resistance to produce motion, and it must continually overcome that resistance to maintain the motion. As Aristotle had it, there must be an excess of the motive power to the resistance. Two very different representations were proposed to quantify this notion of *excess*. One, due to John Philoponus in the sixth century, states that the speed (motion) is determined by the excess of the force taken as a difference to the resistance or $V \sim F - R$. This form was adopted by the Islamic scholar Avempace in the twelfth century and was best known in the Latin west through the commentaries on the *Physics* by Thomas Aquinas, Albertus Magnus, William of Ockham, and Dun Scotus. They also presented the alternative construction, due to Averroes, that translated *excess* as a ratio, that is $V \sim F/R$. While for Avempace, the resistance can be diminished without limit such that the force is needed for motion only because in some interval of time the moving body must traverse a distance, for Averroes it meant the refutation of the vacuum.

The difference solution satisfied Aristotle's requirement that motion is impossible if the two opposing tendencies are equal. It didn't, however, maintain ratio since the ratio of differences isn't a simple constant. Arithmetic examples, simply numbers pulled out of the air but usually simplified by employing multiples of two, argued against the Avempace difference solution because ratios are not preserved by differences. On the other hand, while the ratio solution explained the role of the medium in producing and maintaining motion it had a flaw. If the force equals the resistance, the velocity doesn't vanish.

The theory of proportion is important for understanding both the Mertonian arguments and the later novelty of Galileo's treatment of kinematics and dynamics. For "arithmetic proportion," differences, the quantities being compared must be of the same *quality*, that is type; we would now say they must have the same dimensioned units. For instance, if we subtract two forces from each other, the difference must be a force. But if this applies to the force law of Philoponus and Avempace, then the velocity cannot be simply proportional to the difference between the force and resistance. Instead, forces can be taken in ratio to forces, while velocities can be taken as ratios, and the resulting dimensionless numbers can then be compared. So to insist that Aristotle's view meant a ratio of force to resistance allowed a comparative ratio of velocities. The difficulty is this only vanishes in the limit of zero force, otherwise for any resistance—as Bradwardine and the Mertonians realized—any force no matter how small will produce motion without a threshold.

The continued application of a force maintains constant speed. To produce acceleration requires a change in the force and/or resistance. We can imagine, then, a continually increasing force for a falling body as it nears its natural place after being released at some height. In any interval of time, the velocity increases so the distance traversed in any interval is $\Delta s = \frac{1}{2}V\Delta t$ where the factor of 1/2 comes from the average speed during any interval Δt. But since the velocity increases as the time, this produces the law for continually accelerated motion that the distance increases as the square of the time. Notice this doesn't take into account ether the source of the acceleration or the dimensionality of the physical quantities involved. It simply translates into a quadratically increasing distance covered and, therefore, a harmonic law for the intervals such that for successive identical intervals of time the increments of distance are a simple arithmetic series $(1, 3, 5, \cdots)$. Both of these results would again appear in Galileo in the seventeenth century, while the quantitative aspect was developed both at Merton and Paris.

After dismissing these first two formal laws in the *Tractatus*, differences or simple ratios, Bradwardine settles on one he describes as:

> The proportion of the speeds of motion varies in accordance with the motive to resistive forces, and conversely. Or, to put it another way, which means the same thing: The proportion of the proportions of the motive to resistive powers is equal to the proportion of their respective speeds of motion, and conversely. This is to be understood in the sense of geometric proportionality.

Since Bradwardine, in the first two books, had distinguished the types of proportions, the coda was precise for his readers: the ratio was taken in the form between two, not three, quantities in the sense of force to force, and velocity to velocity. In the subsequent theorems, he elaborates on this with numerical examples that have no physical basis. They're just the usual simple arithmetic choices. Now in the first sections of the work, Bradwardine presented proportions as *arithmetic* (or differences), *geometric* (the ratio of two quantities), or *harmonic* (a special case of geometric proportionality) following the propositions of Euclid, Book 5. Thus, if we take a proportion of proportions of the velocity we mean a number, N, raised to the power of the velocity V such that:

$$N^V = (F/R)^N$$

for an arbitrary N; this satisfies the requirement from Aristotle that if $F = R$, the velocity is zero. Interestingly, there is an anonymous abstract of the *Tractatus de Proportionibus* from the middle of the fourteenth century, not long after the *Tractatus*, that provides a clearer presentation than Bradwardine himself (Clagett 1967):

> The fifth opinion, which is the true one, posits that the velocity of movement follows the geometric proportion of the power of the motor above the power of the thing moved. Whence the meaning of that opinion is this: if the two powers and the two resistances and the proportion between the first power and its resistance is greater than the proportion of the second power and its resistance, the first power will be moved more rapidly than the the second with its, just as one proportion is greater than the other.

which can be rendered, symbolically, as

$$\frac{F_2}{R_2} = \left(\frac{F_1}{R_1}\right)^{V_2/V_1}$$

thus making clear the proportion of proportions part of the formalism and that we are not here comparing incommensurate things. The formula derives from the proportion of proportions since, if we have two quantities $A : B$ and $B : C$, then $A : C = (A : B)^2$, found by dividing both sides by C. Where does this force law come from? Certainly not from any test in our understanding of empiricism. It satisfies all the requirements of dynamics and that's enough.

Geometric reasoning was particularly important to the treatments of this problem and it spawned a new style of reasoning. Mainly founded in the theory of proportion, by equating number (quantity) to attribute (quality), figures—a sort of graphical solution—was developed for representing accelerated motion. It seems to have originated with Nicole Oresme (born around 1323, died 1382) and Giovanni di Casali in the early fourteenth century. Again, this is used to illustrate the text, not as a calculational tool. Uniform quantities are those which can be represented by rectangles, those for which the intensity does not change with extension. A non-uniform quantity, called uniformly difform, is represented by a right triangle, and as Oresme states in *On Configurations of Qualities*,"a difform difform (non-uniformly increasing quantity) is a combination of diagrams. A quality *uniformly difform* is one in which, when any three points [of extension] are taken, that proportion of the distance between the first and second to the distance between the second and third is in the same proportion as the excess in intension of the first over the second to the excess of the second over the third." He continues, "Every measurable thing except number is conceived in the manner of continuous quantity, hence it is necessary for the measure of such a thing to imagine points, lines, and surfaces—or their properties—in which, as Aristotle wishes, measure or proportion is originally found. Here every intension which can be acquired successively is to be imagined by means of a straight line erected perpendicularly on some point or points of the [extendible] space or subject of the intensible thing" (Clagett 1959).

To create a graphical representation, the quality is taken as the vertical axis and the quantity is taken as the horizontal. In other words, a velocity is a quality, the time is the quantity. Then the mean value theorem follows from taking the mean quality (*motus*) as $V_{\text{final}}/2$ and the quantity, time, as Δt so if the distance covered in an interval is $V\Delta t$ it follows that the distance covered varies as $(\Delta t)^2$. Thus a constant quality is pictured as a quadrilateral and every quality that changes uniformly, *uniform difform* starting from zero is represented by a right triangle. Every quality that is *uniformly difform* starting from any arbitrary initial value is a rectangle surmounted by a right triangle. This is a general method that Oresme then applied to motion. Every velocity occurs in an interval of time the duration of which is the base of the figure, or the *longitude*. The quantity or *intension* of velocity is the vertical axis or *latitude*. So figures can be geometrically compared, since they are similar, as ratios of similar quantities (time to time, velocity to velocity), and the distances covered are then also formed from ratios (since the distance is the area of the figure). Even if we are comparing quantities that are not the same (two velocities relative to two times) the graphical construction made it simple to imagine how to do this. This kinematic construction isn't directly related to forces, but it shows how a mathematical scheme was developed in the fourteenth century that could handle accelerations. The rule of proportions was strictly followed, motions were compared to motions, and intervals of space to each other, and intervals of time to each other, all separately. No compound proportions, such as products of velocity and time, were used so this approach doesn't require dimensional specification. Each quality is separately considered.

Uniform velocity, required a different explanation than an accelerated quality. But once the mathematical machinery was in place for describing one type of acceleration, it could deal with any. Swineshead and Oresme knew this well, they both asserted that any variation of accelerated motions can be represented by alterations of the geometric figure. In this, dealing with rectilinear motion, they also know that accelerations are simply additive. The cause of the acceleration was not immediately important, that would come with the Italian mechanicians about two centuries later. This method can be easily extended to include any representation of the change in the magnitude so the Merton rule follows immediately: in computing the interval covered use the mean value. So a speed that is uniformly changed during an interval is the same (in the later language, as a constant acceleration) so the distance covered in any interval of time—starting from the motion—increases as the square of the time. This is a precise kinematical definition, nothing more. It lacks any explicit statement of the *cause* of the acceleration. But, since the underlying assumption is that no motion occurs without a driver, and any change in the motion reflects a change in the driving, the dynamical implication is already built into the discussion. *As long as motion cannot occur except by continuing effort of agents, internal and/or external, that change the qualities, any kinematic discussion is automatically a dynamical treatment.* Without *inertial* motion, the separate treatments were unnecessary. Any change in the driving that produces a variable acceleration can be broken into intervals of constant acceleration. You see already the need to consider infinitesimal intervals of change. Note that this construction is extremely general, any property can be graphed that depends on any interval, it isn't restricted to motion—for example, the ""hotness" of a body depending on the amount of fire in which it is in contact or its extension with respect to an applied weight—so this is a tentative start on quantification. It just wasn't used that way at first.

On the issue of intension and remission of forms, one good example comes from an anonymous piece from the Venice 1505 edition of the *Tractatus*, entitled *Summulus de Moti Incerti Auctoris*:

> 4. It is noteworthy that there are two kinds of intension, quantitative and qualitative. Quantitative intension takes place by adding quantity to the maximum degree of that quantity. Qualitative intension can take place through rarefaction or condensation, e.g. if there were a body one half of which was at a maximum hotness and in the other part the heat was remitted. Then if the more remiss (i.e. less hot) half were condensed, while the other part remained as before, the whole is understood to be

> hotter because the ratio of the more intense to the more remiss part will be continually in greater proportion . . . And just as space is that which is to be acquired by local motion, so latitude is that which is acquired by motion of alteration. And in the same way that a moving body which describes more space in an equal time is moved more swiftly as to the space, so that which by alteration acquires more latitude in a certain time is altered more swiftly.

With 20–20 hindsight, it seems astonishing that the Scholastics made virtually no appeal to experience when proposing these force laws. Some of the simplest demonstrations, such as comparing dropped bodies or looking at trajectories of projectiles, were never reported in the mechanical works. There were a few remarkable exceptions, the English thirteenth century philosophers Grosseteste and Roger Bacon, but their interests were limited mainly to optical and meteorological phenomena and even those were not reported in a way that makes their methods clear. The general procedure was to propose a law through the filter of their presuppositions and use only logic to arrive at a conclusion. In this manner the commentaries were frequently like the illustrations in the manuscripts, reproducing rather than extending the text. Even in astronomical texts, the phenomena were rarely represented from nature, even when tables are included the data are often merely geometrically corrected for the appropriate epoch of the text and not from new observations. The diagrams were intended to guide the reader and, although this wasn't the rule, occasionally drawn by scribes who had little understanding of the material they were copying. When comments are made about everyday experiences they are of a very general sort. The Scholastics, in general, distrusted the senses and lacking empirical principles to guide them eschewed experiment as a means of discovery. The texts themselves were the objects of study and the commentary was the principal tool for explicating the text, citing authorities to expand or contradict the material. But this was about to change.

THE SETTING OF RENAISSANCE PHYSICS

On 23 February 1455, Johannes Gutenberg completed the printing of the "Bible in 42 lines," the first book produced in Europe using movable type. Within 50 years, printers had begun publishing texts in Latin, Greek, and Hebrew throughout Europe with astounding results. At last texts that could only be painstakingly reproduced by professional scribes in single copies were available in tens, or even hundreds, of copies. The efficiency and accuracy of the manufacture, and its comparatively low cost, along with the expansion of centers of paper production, made it possible for the literati in very different, distant places to read exactly the same text. As the reading public expanded so did the demand for new material and the rate of book production rapidly grew along with the range of topics. It shouldn't be thought that producing a book wasn't a difficult, time consuming job, but in the time it took to copy a single manuscript it was possible to obtain dozens of "pulls" of the pages from the press. Not only classics and commentaries issued from the new publishing firms. Poetry, romances, music, histories, maps, and—most important for our discussion—scientific and philosophical works began to appear.

The loci of activity expanded. New schools and universities were being founded but there were also the secular courts of the powerful families, for whom a local savant was a prized "possession" and adornment; perhaps the most obvious examples are the Dukes of Urbino and Ferrara, the Medici of Florence, and Rudolf II of the Hapsburgs. Cities, such as Venice, and royal patrons continued their role of supporting this new intellectual class along with an increasing interest of the ecclesiastical authorities in the exploding scholarly activity (though rarely for the same reasons).

A new figure, the knowledgeable amateur, appeared on the scene. Leonardo da Vinci (who we will discuss presently) is certainly the prime exemplar but there were other artists, such as Albrecht Dürer, Botticelli, Michelangelo, and Raphael who were fascinated with the new ideas of nature and the abstractions of space and time. But there were also those who had the wealth and leisure to engage in these activities full-time, Tycho Brahe and Ferdinand II of the Medici family are among the most famous. These all now had access to a wider dissemination of their ideas through the press and in some cases, such as Tycho's, their works were fundamental influences in the debates of the scholars.

An increasingly complex set of scientific instruments also became available in this period, produced by skilled specialists. More advanced glassblowing techniques, for instance, not only benefited alchemical studies but also physics, the barometer and thermometer are examples. Clockwork and gearing, and astronomical instruments (not only the telescope at the start of the seventeenth century but also astrolabes, quadrants, compasses, and geometric instruments) improved the accuracy of measurements. On a darker side, the introduction of gunpowder and the increasing use of ballistics in warfare required increasingly more precise knowledge of dynamics and materials. An archer required only skill to be effective, having complete control of the tension and aiming of the arrow and bow. To hit a distant target taking full advantage of the capabilities of the new artillery required increasingly precise methods for computing trajectories and the right quantity of explosive propellant. Rules of thumb went only so far, understanding their origins became a major theoretical occupation that continued for many centuries (even now). These developments were not only in the area of dynamics but also statics, since fortifications needed to withstand new perils. The doctrine of impetus became increasingly useless for computing quantitatively ballistic motion and, as we will see, even called into question some of the foundational assumptions of physics.

The great architectural projects of the Renaissance, inspired by the classical works of Vitruvius and Hero, provoked a new interest in the abstract study of physical behavior of materials. Grand projects are costly and the exigence of the time was originality. It wasn't enough for a master builder to slightly modify an existing, successful design by scaling it up or down to the demands of a new contract. The enormity of some of the new structures, for instance Brunelleschi's dome for the Santa Maria del Fiori cathedral of Florence and Michelangelo's and Della Porta's for St. Peter's basilica in Rome, required an advanced understanding of materials and stresses. While many of the studies, those of Leonardo being the most detailed surviving examples, were mainly qualitative, the phenomena were being carefully examined with an eye toward the underlying principles and causes. This was the environment in which science entered a new phase, in which the empirical and theoretical, joined in the language of mathematics, produced a new physics.

EXPERIENCE AS GUIDE: LEONARDO DA VINCI

Leonardo (1452–1519) was a bridge between two epochs in many aspects of western culture. His contemporary reputation rested on a comparatively limited—but remarkable—artistic production, a few disciples of outstanding quality, and his notable technical expertise. Our usual categories, however, fail properly to describe his activities. He was an itinerant artist working in the employ of several epochal patrons in Italy and, eventually, in the French royal court. He was an engineer—really a technical consultant—to the Sforzas (Milan) and Medici (Florence). Although none of his grandiose projects were ultimately realized, he did have a significant influence on later developments, especially in fortification and urban planning. His reputation as an empiricist and precursor of developments in the physical and biological sciences comes from scholarly studies that have mined his scattered literary productions since the late eighteenth century.

Leonardo is important for our discussion of physical ideas in the late middle ages precisely because his results were arrived at and preserved in such private circumstances. His notebooks record the musings of an amateur. He held neither an academic nor court position, nor was he a theologian or natural philosopher. If we are to judge him by his own standards, we can use a letter to Sforza, the Duke of Milan, as his "curriculum vitae" for possible employment in the service of the Duke. He writes of his skills as an architect and practical mechanician, expert in military construction and ballistics, hydraulics, and urban planning. What we find in his notebooks, in contrast to this presentation, is a brilliantly observant, intensely curious, outstandingly empirical individual who frequently, and with erratic motivations and methods, addresses often profound problems with a very personal mythology and perspective. It is precisely because some of these musings, often queries without significant elaboration, but also long connected fragmentary gropings along the way to the resolution of a particular question, that Leonardo has often been presented as a modern. This is often the case in his nature studies, which have the stamp of a biologist and systematist. But Leonardo's physics show that he was an Aristotelian in his theoretical dynamics, even though he was often inconsistent in his views. He was above all else an applied scientist who established the concept of copying from nature by abstraction. In this, he achieved what Bacon and Grossteste had advocated to a less sympathetic audience three centuries earlier: that experimental science and observation can test knowledge and correct first principles. His work shows an acquaintance with mathematics but not with formal geometry, so his ideas often lack the precision Archimedes' analysis. Nevertheless, Leonardo was able to extract some fundamental concepts from relatively simple observations, often of things in their natural settings. For example, his study *On the Flight of Birds* represents a founding work in biomechanics and, consequently, in the application of the force concept to the analysis of living things. Here Leonardo's insights far outstripped his abilities to formalize and extend them. His scaling arguments are based on simple ideas of lift not associated with the movement of a body through air but just on an extension of the support by atmospheric pressure. His studies of turbulence are outstanding and are relevant to our discussion because they provide the first systematic analysis of the *motions* of a continuous medium. Descartes used remarkably similar ideas in his theory of planetary motion without any knowledge of his forbearer.

Leonardo's extensive studies of statics bear greater similarity those of Hero of Alexandria than Archimedes. His aims were generally practical although his observations often go beyond experience to attempt to extract general principles. In his description of weights suspended by cords of unequal lengths, he qualitatively described the law of addition of forces. How does a force alter a body? How does it achieve balance when there are different internal and environmental influences? Questions of this sort were particularly compelling for Leonardo and to address them he extended the mechanistic tradition of Archimedes and Hero. For practical purposes, he was also especially interested in how materials behave under stressing and loading. His discussion of positional gravity was along the lines pioneered by Jordanus but when it came to extended bodies he went farther. His description of the differential reaction of a stressed beam displays his originality and illustrates his way of reasoning about mechanical problems:

> If a straight spring is bent, it is necessary that its convex part become thinner and its concave part thicker. This modification is pyramidal and consequently here will never be a change in the middle of the spring. You shall discover, if you consider all the aforementioned modifications, that by taking part *ab* in the middle of its length and then bending the spring in the way that two parallel lines *ab* touch at the bottom the distance between the parallel lines has grown as much at the top as it has diminished at the bottom. Therefore, the center of its height has become much like a balance for the two sides. (Codex Madrid, trans. Carlo Zammattio)

With this remarkable statement, Leonardo empirically anticipated a fundamental aspect of deformation that appears to have been ignored by all earlier Western writers and would not appear again until the eighteenth century. I'm emphasizing this as a striking example of how relatively simple the reasoning could be when physical problems were stated geometrically and how notable it is that nobody noticed something that Leonardo considered almost obvious. We will return to this in more detail when we discuss the developments in statics after Newton.[4] But Leonardo's dynamics was, in contrast, consistently expressed in the framework of the impetus description for and explanation of ballistic motion. He never systematized his observations in geometry, but his jottings show that he viewed "violent" motion as a continuous variation between a curved trajectory and freefall. The flight of a cannonball was, for instance, a continuous curve that gradually lost impetus and was replaced by natural motion in an asymmetric arc.

NOTES

1. The *Method* was only recovered at the very end of the nineteenth century when Heiberg, one of the foremost scholars of ancient Greek mathematics, recognized the presence of the text *under* another theological text, virtually scrapped of the parchment to reuse the precious material. Its final recovery, transcription, and publication came more than 2000 years after its composition. Such was the fate of many of the works of the Greeks

and Romans for which we know only the titles and abstracts of the contents from later commentators, and not only in Europe.

2. It is rare that history can be re-lived as well as recovered but there is actually an interesting analogy to the Medieval reencounter with the ancient texts taking place in real time. The medieval city Timbuktu, now in Mali, was one of the principal trading posts along the Niger in the twelfth century. The cache of manuscript material, randomly collected there, has yet to be explored. While it is unlikely that any great number of previously unknown texts will be discovered amid the manuscript material, it is something you can watch develop with its historical precedent in mind.

3. I think an aside is in order here since so much of the technique used in the fourteenth century discussions has been preserved in the pedagogy of modern introductory physics textbooks. To a modern reader, so accustomed to nearly instantaneous communication, mere chronology doesn't, I think, make clear how hard it was to pass from the subtle disputes of the Schoolmen to the first dynamical studies of Tartaglia and his Italian contemporaries. Keep in mind, almost *200 years* separated the major works of the Mertonian and Parisian schools and the Northern Italians. And it would take nearly 60 years after that before the concept of force would be harmonized with these kinematic, or semi-kinematic, treatises.

4. Strings, and springs can be treated in one dimension, at least initially. But when you have a beam, or a membrane, or a plate, the problem is compounded by the multidimensional nature, and multidirectional character, of the surface and forces acting on it. For example, a string can be described by a linear relationship between the applied stress (force, loading, weight) and the reaction strain: it changes in length and the more stress is applied the greater the distension. This is the first form of the law described by Hooke (1672) and Mariotte (1680) for the linear dependence of the deformation (strain) to the applied stress. The innovative point is that the relation is a simple proportion, with the constant of proportionality being a specific only to the material and not to the deformation. This linear law, the truth of which is restricted to a limited range of stresses and materials, was a step far beyond the rheology available to Galileo.

3

THE NEW PHYSICS

> Let sea-discoverers to new worlds have gone,
> Let Maps to other, worlds on worlds have shown,
> Let us possess one world, each hath one, and is one.
>
> —John Donne, *The Good-morrow*

The span of two centuries between c.1500 and c.1700 marks the period usually called the *Scientific Revolution*, between the appearance of the heliocentric system of Copernicus to the age of Newton, neither is a firm endpoint. As we've seen, Aristotelianism was deeply rooted in the cultural soil and held firm for a very long time, throughout this period. But at the same time there was a new, growing sense of the importance of actual observation and experimentation as a way to discover regularities of nature. Gradually the intellectual climate shifted from one depending on authority, the more antique the better, to individual experience. Instead of being content with repeating, commenting on, or criticizing ancient authorities, it was an age of hypotheses. It opened with an epochal event: the first European encounter with the Americas. It was a disruptive shock. Suddenly the phenomenology of the world was populated with peoples and species of plants and animals never before seen. The categories no longer held so simply as they had in the Old World, and the exploitation of these new discoveries required new ideas and methods. As warfare had transformed during the previous century, now commerce changed. The world wasn't simply a sphere, that was already well known of course long before Columbus and the voyagers. It was far larger, far more diverse, and much less familiar after the beginning of the sixteenth century. It was an age of discovery.

HELIOCENTRICITY: COPERNICUS

The separation between cosmology, that is the physical explanation of the planetary motions, and astronomy, which dealt mainly with their description (and astrology,

NICOLAI COPERNICI

acto in c ſemicirculo, apparebit Sol Cancrum ingredi. At F auſtrina æquinoctialis circuli declinatio ad Solem conuerſa, faciet illum Boreũ uideri æſtiuum, tropicum percurrentem pro ratione anguli E C F inclinationis. Rurſus auertente ſe F ad tertiũ circuli quadrantem, ſectio communis G I in lineam E D cadet denuo, unde Sol in Libra ſpectatus, uidebitur Autumni æquinoctiũ confeciſſe. Ac deinceps eodem proceſſu H F paulatim ad Solem ſe cõuertens, redire faciet ea quæ in principio unde digredi

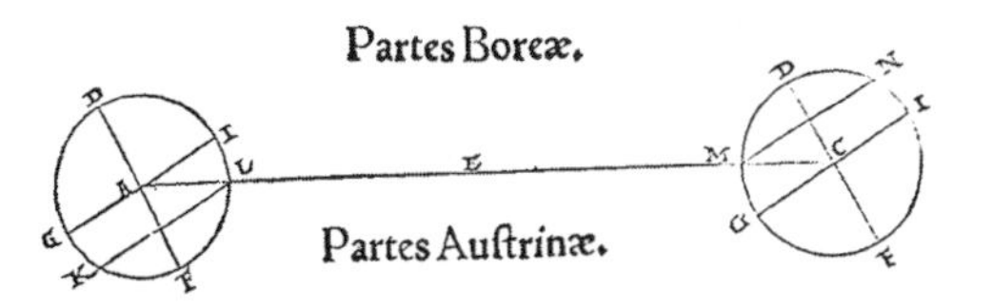

cœpimus: Aliter. Sit itidem in ſubiecto plano A E C dimetiens, & ſectio communis circuli erecti ad ipſum planum. In quo circa A & C, hoc eſt ſub Cancro & Capricorno deſignetur per uices circulus terræ per polos, qui ſit D G F I, & axis terræ ſit D F: Boreus polus D, Auſtrinus F, & G I dimetiens circuli æquinoctialis. Quando igitur F ad Solem ſe conuertit, qui ſit circa E, atq; æquinoctialis circuli inclinatio borea ſecundum angulum, qui ſub I A E, tunc motus circa axem deſcribet parallelũ æquinoctiali Auſtrinum ſecundum dimetientem K L, & diſtantiam L I tropicum Capricorni in Sole apparentem. Siue ut rectius dicam: Motus ille circa axem ad uiſum A E ſuperficiem inſumit conicam, in centro terræ habentem faſtigium, baſim uero circulum æquinoctiali parallelum, in oppoſito quoq; ſigno C omnia pari modo eueniunt, ſed conuerſa. Patet igitur quomodo occurrentes inuicem bini motus, centri inquam, & inclinationis, cogunt axem terræ in eodem libramento manere, ac poſitione conſimili, & apparere omnia, quaſi ſint ſolares motus. Dicebamus autem centri & declinationis annuas reuolutiones propemodum eſſe æquales, quoniam ſi ad amuſſim id eſſet, oporteret æquinoctialia, ſolſticialiaq; puncta, ac totam ſigniferi obliquitatem ſub ſtellarum fixarum ſphæra, haud quaquam permutari: ſed cum modica ſit differen-

Figure 3.1: The illustration from *de Revolutionibus* showing the explanation for the seasons related to the constancy of the direction in space of the rotational axis of the Earth during its annual orbit. Image copyright History of Science Collections, University of Oklahoma Libraries.

treating their interpretation), allowed for a broad latitude in any mathematical kinematics. It also produced striking contradictions. We've already seen that epicyclic motion violated at least one principal that any body should have a single natural motion. In Ptolemy's system, with both the epicycle and the equant display compound natural motion about centers other than the Earth. Yet the system was made at least minimally consistent by imposing uniform motion of the deferent around the center of the world, even though it was eccentric. In contrast, the Pythagorians asserted that the Earth moves around the central fire of the Sun, though without developing either a full scale physical picture or a predictive geometric apparatus. The system survived as an almost mystical tradition for a very long time, even long after the critiques by Aristotle, and was best known through those citations since the adherents deigned to write anything substantial. The most dramatic departures from Aristotelian cosmology came not from the physical but the mathematical side, the hypothesis by Aristarchus of a moving Earth. None of the physical consequences of this hypothesis were examined, as far as we know, but its evident simplification of the motions exerted an enduring influence. This is especially interesting because, aside from his work on the distance to the Sun, Aristarchus' cosmological construction opinions were only known through a brief mention in Archimedes' work *the Sand Reckoner*. For our purposes, it represented more of a model representation of, than a physical precursor to, heliocentricity.

If we reserve the term "solar system" for the complete model that combines the physical and mathematical principles in a self-consistent way, we have to wait nearly two more centuries, but the process began dramatically with Nicholas Copernicus (1473–1543). From his first sketch of the idea, the *Commentariolis* of 1513, to the publication of his crowning achievement, *de Revolutionibus Orbium Coelestium*, took 30 years and he never lived to see its impact. The book's principal arguments were popularized by his pupil, Georg Rheticus (1514–1576), who wrote a summary, the *Narratio Prima*, that circulated separately as sort of a lecture on the main ideas. What Copernicus had done was more than merely making the Earth move. He saw what this implied in the big picture, that previously diverse and unexplained phenomena could be included naturally within the world system and that, in consequence, new physics would be needed to account for the motions. But he did not live long enough, nor have the necessary physical concepts at hand, to take that last step. For Copernicus, celestial *dynamics*, the mechanical part of the work, remained largely unaltered from those of the Hellenistic system of the spheres. A driver beyond the system, no matter how much enlarged, was still required and the action was transmitted to the planets by the spheres. What changed was that this outer driver didn't need to participate in the motion of the plants. The rotation of the Earth, not of the eighth sphere, explained all of the motions of the fixed stars, whether daily or the precession of the equinoxes, by "reflection." The Moon moved with the Earth, somehow. These were more than hypotheses of convenience, for Copernicus they were real. But although the center of this world had been displaced from the Earth to the Sun, there was little change in the construction by the humanist scholar Georg von Peurbach almost 70 years before in his *Theorica Nova Planetarum*, in which the planets ride like ball bearings between two rigid spheres in what looks like a mechanical transmission. But for von Peurbach and his contemporaries, except Copernicus, each mover required a different construction centered separately on the Earth. This is the sweeping innovation of *de Revolutionibus*: the system was unified and it avoided many of the paradoxes of the geocentric system, such as mutually intersecting epicycles and unphysical changes in the angular diameters of the Sun and Moon. To account for deviations in the orbital rates, planets were carried on small epicycles, and the effect of the equant was replaced by additional small epicycles to maintain the constancy of local circular motion (that is, motion was always assumed to be constant around the center of the path, eliminating both the eccentric and the equant). The truly remarkable elements of Copernican heliocentrism were the requirement of the multiple motions of the Earth and Moon and the conception of a force not centered on the Earth. Although eschewing any explicit discussion of dynamics, it is clear that Copernicus' views require a fundamental reassessment of Aristotelian physics, so much so that Andreas Osiander, who edited the first edition and guided the work through the press, felt compelled to add an anonymous preface in which the system is called only a "hypothesis" from which no physical implications *should* be drawn.[1] That Copernicus himself understood those implications is, however, not in doubt. Following the sequence of the arguments from the *Almagest*, Copernicus presented in his first book a detailed description of the origin of the seasons—due to the immobility of the Earth's rotational poles, and a mechanism producing precession of the equinoxes. The annual motion and fixity of the poles is very important for the force concept. It implicitly uses, for the first time, what would later be called conservation of angular momentum, although this idea isn't found in *de Revolutionibus*. The other, fundamental, feature of the system is that not only were bodies endowed with multiple *natural* motions, but one of them is the rotation of the Earth.

Epicyclic motion is a produce of the frame of the observer and is, in effect, a choice. If viewed in the stationary frame centered on the Sun, planetary motion is always direct and periodic. In a rigidly rotating system, one for which all bodies have the same orbital period, all frames are kinematically equivalent. But if the periods change with distance, motion must reverse for some of the bodies depending on your choice of point of observation (that is, if you are on one of the moving bodies). This was understood by Copernicus but most clearly explained by Galileo (in the *Dialogs on the Two Chief World Systems*, almost a century later) and exploited by Kepler (in his *Epitome of Copernican Astronomy*). The stationary points result from those intervals when the planets pass each other and, from the viewpoint of the more rapidly moving body, the slower is overtaken. Then the confinement of the motion to closed orbits produces the strict periodicity in these with the separation of the times of the event being dictated by the difference in the frequencies of the two orbits.

In a broad sense Copernican kinematics was simpler than classical models but not entirely free of its constructions. To account for the irregularities in the periodic motions (the principal anomaly), Copernicus still required small epicycles that would produce the necessary deviations in the rates. Remember, there were no absolute distance scales so there was still no way to connect the motions with terrestrial forces. The motion of the inner planets, which we can now properly name because of the change in the reference frame, found an immediate explanation with the heliocentric construction: Mercury and Venus are slaved to the solar motion because of their relative distances from the Sun and their location interior to the Earth's orbit. The relative distance scale, that was lacking in the *Almagest* and all subsequent work, was introduced, at least for the inner planets. Using purely geometric arguments, the sizes of the orbits relative to the Earth's could be determined since this depends only on their maximum angular separation from the Sun.

A mobile Earth removed the greatest cosmological impediment remaining from the Aristotelian system. Now the world could be infinite, the stars were no longer fixed to a sphere that transmitted the force of the Prime Mover to the planetary machinery. A new source was needed for the motions.

In the dynamics of the spheres, since the construction was designed to "preserve the appearances," each motion was identified and treated separately. The mean motion that defined the circular path around the ecliptic was a reference motion but each departure from that referent was treated separately using the same or a related mechanism. The complication of the model was irrelevant, its success was enough to guarantee its adoption. All the various bells and whistles tacked on to the fundamental construction, the eccentric, equant, moving equant, and the like, were all required alterations whose physical origins were deemed beyond the scope of the approach. The simple necessity of accounting for a departure from a successively more complex reference motion was the product of necessity and the mathematical tractability of the picture was enough to recommend the procedure.

For Copernicus this was still a valid method. With the displacement of the center of motion from the Earth to the Sun came new questions but not a real change in the physics underlying the machine. It was still a *machine* in the sense that its parts were truly driving mechanisms that maintained the structure and kinematics. It may now be hard to imagine, at a distance of so many centuries, how effectively this separation could be maintained. Even in the face of mounting contradictions in the hypotheses regarding the dynamics, the *calculations* of planetary positions could be affected without ever asking such questions. Those who chose to avoid asking cosmological questions were able to do so without having to ally themselves to a physical explanation of the system. For Tycho and those who attempted to maintain some point of contact between the physical causes and the observed effects, this was the weak point of the Copernican approach but nonetheless it was still a matter of almost personal preference which model one chose for a geometrical arrangement of the principals.

The physical contradictions were there in abundance, however, and even Copernicus was aware of some of the consequences of this altered construction. He never, however, addressed them in a systematic way, focusing instead on the advantages of the new ordering. For Rheticus this was also acceptable, as it was for other Copernicans. Actually, the theological and philosophical consequences of the new system were far more disturbing than their physical implications. The motion of the Earth was, perhaps, the one sticking point. The arguments adduced by Ptolemy and in the Middle Ages for the stationarity of the Earth were so obvious and simple that they sufficed to persuade most casual critics that the system could function as a simple change in computational reference frame but had no other significance. For the Copernicans, the physical reality of the system—as opposed to its mathematical convenience or efficiency—was more dramatic. It *must* be possible to decide by experience if the Earth is *really* moving or not. If it isn't, the heavens are in one way far simpler, the universe is a closed thing and God can be in His heaven. If, on the other side, the Earth is mobile, then this whole construction fails to stay together *in a unified sense*. You could say it was the happy mistake that the clerics of the Renaissance chose to base their theology on physical demonstration rather than simply using faith and morality. But they so chose and this opened the need to keep the whole coherent.

EXPERIMENT AND OBSERVATION: WILLIAM GILBERT AND TYCHO BRAHE

In 1600, William Gilbert (1544–1603) authored the first systematic experimental analysis of a force. In *de Magnete* ("On the Magnetic"), he presented detailed accounts of an enormous variety of experimental studies he conducted on the lodestone, the naturally occurring iron mineral material from which magnet is were originally made. As a physician—he was the court attendant to Elizabeth I—Gilbert was schooled in Galenism with its language of "humors," "efluences," and "affections" and he used this language to describes the phenomena he observed. In the work, he never mention force, or even impetus, but his affinities and affections—the same terms used by the alchemists of the period—are similar in intent if not in detail.

Gilbert's method was purely empirical, drawing his "hypotheses" solely from experiment. He was the first use a material model as a representation of a physical body, in this case leading to his principal contribution: the realization that the Earth is a magnetic with its source within the body. This allowed him to use a compass needle, placed alongside and moved around a spherical lodestone, to simulate the behavior of the then common mariner's instrument. His explanations of the dip (declination) and orientation (deviation) of the needle as a function of

76 GVILIEL. GILBERTI

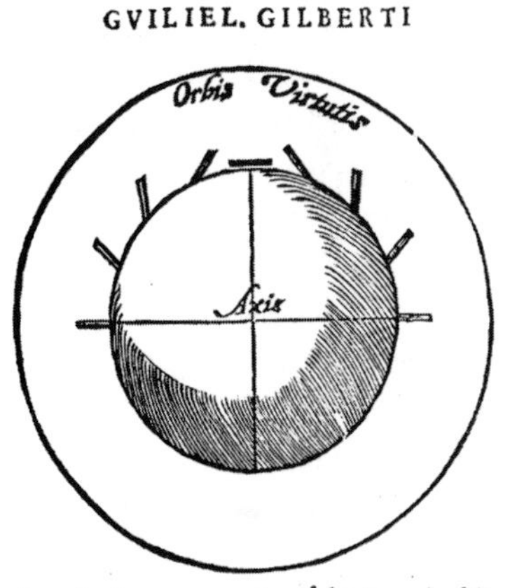

Quò propiores fuerint partes æquinoctiali, eò magis obliquè alliciunt magnetica: at polis viciniores partes magis directè aduocant, in polis directissimè. Eadem etiã ratio est conuersionis magnetũ omnium qui sunt rotundi & qui sunt longi, sed in longis experimentum est facilius. Nam in quâuis formâ est verticitas, & sunt poli; sed propter malam formam & inæqualem, sæpiùs quibusdam malis impediuntur. Si lapis longus fuerit, vertex verò in finibus, non in lateribus; fortiùs in vertice allicit. Conferunt enim partes vires fortiores in polum rectis lineis, quàm obliquis. Sic lapis, & tellus naturâ conformant motus magneticos.

CAP. VII.

De potentiâ virtutis magneticæ, & naturâ in orbem extensibili.

Funditur virtus magnetica vndequaque circa corpus magneticum in orbem; circa terrellam sphæricè; in alijs lapidum figuris, magis confusè & inæqualiter. Nec tamen in rerum natura subsistit orbis, aut virtus per aërem fusa permanens, aut essentialis; sed magnes

Figure 3.2: A figure from Gilbert's *De Magnete* showing the "influence" of the magnetism of a lodestone in the form of a sphere, the dip of a magnetic needle. This model of the Earth's magnetic field, the terrela, provided the empirical basis for later attempts in the seventeenth century (notably by Edmund Halley) to determine longitude with magnetic measurements. Image copyright History of Science Collections, University of Oklahoma Libraries.

longitude and latitude provided the theoretical basis for navigation and, later in the work of Edmund Halley, produced the first means for determining longitude independent of celestial measurements.

Gilbert was a forceful advocate for the Copernican system. In the last book of *de Magnete* he presented defense of the heliocentric system but went even farther into the physical consequences than had Copernicus. He takes two enormous leaps, both extensions of his *magnetic doctrine*. One is the banishment of the spheres. Tycho Brahe had done the same in his modified geocentric system. In their place, Gilbet asserted that the Sun is the source for the motion of the Earth and the planets through a sort of magnetic influence. The same idea would be proposed, in a more physical way, by Kepler. Gravity is something distinct and different for Gilbert as for all others, the action of a magnetic is something almost "voluntary" on the part of the substance and not all objects feel the influence of the lodestone. The other is a much more wide-ranging idea, that the stars are not only distant but at a vast range of distances, and that there are innumerably many fainter than the limits of vision. While he leaves this unexplored and draws no other implication from it, for our purposes this hypothesis is a suggestion that the open universe, which Newton would need for his system, was already being considered a century before the *Principia* and a decade before the Galileo announced its reality in the *Sidereus Nuncius*. In the final analysis, Gilbert's book represents a bridge rather than a starting point. His experiments were reported in almost mystical terms.

Gilbert wasn't alone in his phenomenological approach. A fellow traveler along this empirical road, the Danish nobleman Tycho Brahe (1546–1601), proposed a significantly modified—but still essentially Aristotelian—system of the world at the end of the sixteenth century. His attempt was to reconcile two contradictory exigencies. One arose from his astronomical observations, especially of the comet of 1577. Finding that this body was superlunary, in showing no detectable parallax, Tycho's construction of its orbit definitively required that it cross the boundaries of the spheres. Thus, absent now the principal driving mechanism and the basis for the previous unification of the planetary motions, he proposed a radical non-Copernican alternative to the Ptolemaic conception, almost 50 years after the appearance of *de Revolutionibus*. Tycho's systematic observations of planetary motion and stellar positions, and his determination of the motion of comets and their implication for the driving of the planets, were important for the reform of astronomy but ultimately secondary to the development of physical theory. Like Gilbert and others of the sixteenth century, he was content to "preserve the appearances" without mechanism. His planetary theory is a mathematical construction, still couched in vaguely Aristotelian terms—the immobility of the Earth is assumed—but with a bizarre arrangement of the planetary order to account for the motions. The Earth is immobile in this system, the planets circulate around the Sun, but the Sun moves around the Earth along with—and above—the Moon. His system is purely kinematic once the motion is constrained to remain geocentric. It can be nothing else since he dispensed with the driving spheres and, lacking any alternative explanation for the motions, was forced to present the system as a mathematical hypothesis without other justification. Although it was sufficiently novel to attract some adherents. Even after Galileo (as late as the mid-seventeenth century), Riccioli and his fellow Jesuits found it more acceptable than the Copernican alternative since it was capable of producing positional predictions while not conflicting with the theologically proscribed geocentricity. But for our narrative it remained just a sidelight.

Gilbert and Tycho represent transitional figures in the history of physical science. Both recognized the limitations of their predecessor's methods and conclusions, and they were convinced that quantitative empiricism was the only way to study nature, but neither was able to break with the principles they inherited from their medieval roots. At the same moment Tycho demolished the crystalline spheres, he sought to maintain a geocentric world and Aristotelian dynamics. Even as Gilbert discovered the properties of magnetic and electrical substances with superb experiments, his speculations invoked Galenic mechanisms of humors and influences. Tycho could maintain a purely geometric construction that violated even his physical axioms and swing between descriptions of the planetary motions and terrestrial phenomena without a hint of admitting their basic contradictions, maintaining an implicitly agnostic stance in lieu of a unified system. Gilbert could create specially tailored forces to account for phenomena without asking about their more general applications. Thus, at the opening of the seventeenth century, although the earlier worldview was showing fatigue, it had yet to crack. That required a new approach, a new openness, and a greater willingness to trust experiments and mathematical reasoning.

KEPLER: ASTRONOMICAL PHYSICS AND NEW PRINCIPLES

The Copernican system, as a mathematical construction, was more than tolerated. It was *argued*, it formed the intellectual fodder for countless debates and contrasts between calculational systems for nearly two centuries. But the question of which method—given that by the start of the seventeenth century there were three available—provided the best means for computing planetary phenomena is not what we're concerned with here. If we are talking about *physics*, the picture is quite different. Many of the debates were cosmological—that is, philosophical or metaphysical—and they centered on the *mechanism(s)* implied by the choice of system. What basic physical picture Copernicus pictured, we can only hypothesize. But he had made a spectacular leap of imagination that would drive many more in the centuries ahead, one that is much different than just a mathematical choice of reference frame: without justification, raised to the level of a postulate, the motion of the Earth is used to examine all observable phenomena of planetary motion independent of explanation. To drive the system may have required a system of spheres as complex as the medieval cosmos, but that's not important. By avoiding explicit reference to the *origin* of motion, and instead concentrating on its *manifestations*, Copernicus had evaded the most serious criticism that would have diminished or even removed any credibility in his system. In eschewing dynamical hypotheses, the construction could stand on its own and, if successful in reproducing the motions of the planets, inspire research programs to deal with its physical origin. Nothing in *de Revolutionibus* , however, leaves any doubt about the nature of the Copernican program. The most direct statement is contained in the *Commentariolis*, a short work from around 1514 that provides a precis of the larger research project. There the assumptions are written in unelaborated style and set out at the start of the long work that ultimately produced Copernicus' monument. Whatever philosophical, theological, or personal motivations Copernicus may have had for placing the Sun as the immobile center of the system are unimportant. He had changed the construction of the world and this required a new mechanism. That was Kepler's program, to complete the *solar system*.

We encounter in this new science a new personality, the mathematical physicist, and none is a better example than Johannes Kepler (1571–1630). As an astronomer and mathematician, and above all a Copernican, he took the physical speculations of Gilbert and a remarkable knowledge of observational astronomy from Tycho and forged a new way of unifying physics, a celestial mechanics. This he achieved despite having initially been an adherent of Tycho's system, and the inheritor of Tycho's observational data compilation of planetary motion. Kepler was the first to demonstrate that the Copernican world system actually leads to new laws, his three laws of planetary motion. The result could not have been obtained without admitting the motion of the Earth. Copernicus knew that a change in the reference system from the Earth to Sun centered coordinates removed a temporal alias in the planetary positions, the effect of the annual motion of the Earth. Thus the periods of the planets were now systematically increasing with distance from the Sun. He also knew that he no longer required the extraordinarily complex, multiple motions for the planets, the necessity for the equant vanished, but he retained a small geometric crutch, a sort of small epicycle, to produce the proper variation in the angular motions. This was only a mathematical device lacking any further refinement.

What Kepler found was a set of geometric and *dynamical* laws. The first is that the areas swept out by a line extending between the planet and the sun are equal for equal intervals of time. Another way of stating this is when the planet is closer to the Sun it moves more rapidly. This held not only for the projected angular motions as they had been for all previous theories, *the physical speed varies depending on the distance*; the changes are not merely uniform circular motion viewed from a different vantage point. Thus, if r is the radius line and v is the angular speed (that is at an moment tangent to the orbit), rv is a constant. The second is that the planets orbits are the closed conic sections, circles and ellipse, with the Sun remaining immobile at one focus and controlling the motion, the opposite focus is empty. Kepler overcame his initial reluctance to accept a noncircular path only because of what he saw as the overwhelming observational constraints derived from Tycho's data. The last, which is called the Harmonic Law, is that the square of the orbital period of a planet increase as the cube of its mean distance from the Sun. The statement of this law is actually an implicit proportion since the units Kepler employed, years and the astronomical unit—the mean Earth–Sun distance—actually scale all orbits to the Earth's. Note that all of these laws are the products of the change in reference frame: without the mobile Earth, there is no basis for assigning distance nor is the periodicity understandable.

Kepler's dynamical explanation for the laws is uniquely his: the Sun, as the core of the planetary system, is also the source of its motion. But it is at this point that Kepler's dynamics echo some of the earlier doctrines. Orbital motion is instantaneously tangential. To continue, according to the construction in the *Epitome* and *Astronomia Nova*, this motion is always driven. Having no spheres to serve this role, Kepler located the source in the center—the Sun—and the means with the only force at a distance he knew—magnetism. Unknown at the time, and not directly measured for three centuries after its first invocation by Kepler, this presumptive solar magnetism passes through the planets directly from the Sun and acts according to their mass, their reaction or resistance. Using the Mertonian notation that $V \sim F/R$, for the velocity to vary inversely with the distance requires a force that does the same.

Kepler took a much more far-reaching step with his astronomy than simply finding regularities in the motions of the planets. In extending the Tychonian program of a universe without the spheres and discovering the correct form of the orbits, he felt compelled to apply the new understanding of long range actions of forces in nature—particularly magnetism—to create a new celestial mechanics. It isn't important that the specific explanation proved unsuccessful for many reasons, or even that he couldn't completely break from the earlier dynamical principles. Kepler looms very large because he did something that no one had before: he took a new cosmological system and attached a physical cause to each part in a way that sought a universal explanation for cosmic phenomena concordant with everyday experience.

GALILEO

The Oxford Calculators and their Parisian contemporaries had debated their physics in isolation from experience. There were some exceptions, some natural phenomena (meteorological and geological, for example), and in optics and astronomy where it was almost unavoidable. But despite the polemics of Occam, Grosseteste, and Roger Bacon, any discussion of motion and forces was usually framed as logical problems and abstractions. But during the fifteenth and sixteenth centuries, spurred in part by the new exigencies of ballistics and machines and a

Figure 3.3: Galileo Image copyright History of Science Collections, University of Oklahoma Libraries.

growing influence of humanistic formulations of nature, experience and experiment gradually moved to center stage.

Accelerated motion is the fundamental concept on which all dynamics was subsequently built. We've seen that this was a question that occupied the later Middle Ages. But by the seventeenth century the increasingly numerous contradictions with experience were becoming more anomalous and less easily avoided. The Aristotelian method of requiring causes fit a dialectic approach. The new approach was willing to look at the systematics without requiring consistency with a specific doctrinal principles. You'll recall that the cessation of motion of terrestrial things wasn't really a problem for the Mertonians or their predecessors. In this imperfect world, there are always things resisting the motion so it would seem a simple thing to imagine that any object requires a deliberate act to keep it moving. For a cart, or a river, or a bird this seems reasonable. But there was another, much more serious problem that this picture simply can't explain: motion when you aren't in direct contact with the object and this was more evident than ever in the ballistics of the early Renaissance. There are regularities in projectile motion that seem to be the same as freefall.

Enter Galileo Galilei (1564–1642), born at the peak of the Tuscan renaissance. His education at the University of Pisa began in medicine in a faculty already deeply committed to the scientific reform of anatomy and physiology; Vesalius had delivered lectures there and the university had founded the first botanical garden for the study of pharmaceuticals. But there was little interest in other areas of natural philosophy and mathematics, to which Galileo was drawn, and he withdrew from the formal study before completing his training. Instead, he pursued private study in his preferred subjects, achieving sufficient mastery to obtain a professorship in geometry at Pisa in 1587. He stayed there for four litigious years, years before obtaining the chair in mathematics at Padova, which was then part of the Venetian Republic. He remained for nearly two decades, taking private students, lecturing, and advising the state on various technical matters. Many features of his career resemble his fellow Tuscan Leonardo, then known only as an artist and dabbler but whose investigations in physical science were otherwise unknown. The comparison is apt. Galileo studied many of the same problems and had the same broad range of interests. But he was a very public figure in his work, a polemicist of talent and erudition, and above all systematic and a superb mathematician. His empirical studies completed the Renaissance project that began with Gilbert and Tycho and the mathematical development that began with Kepler without any whiff of mysticism. This involvement as a consultant, during his Padova years, parallels Galileo's development as a natural philosopher with Archimedes. Both had a gift for seeing profound general principles in mundane problems and the concentration needed to develop the ideas, often resulting in practical mechanisms and proposals.[2]

Galileo's Dynamics

During the time in Pisa, and extending through his Padova years, Galileo lectured on physical natural philosophy. It's clear from his earliest works—for instance, the unpublished *de Motu* ("On Motion") from around 1590—that he knew well the various Scholastic force laws proposed in the previous centuries and, with them, the Mertonian rule for calculating accelerated change. But unlike the Scholastics, he concentrated uniquely on the motion and equilibrium of physical systems. Jordanus' concept of positional weight had been applied only to statics and then only using logical arguments. Galileo, instead, conducted experiments with inclined planes, balances, and moving bodies to discover laws and test conjectures.

Why didn't anyone else try this? Even Leonardo, that dispersive inquisitor, never systematically explored the dynamics of gravitational motion. And nobody had realized that the motion is independent of the mass of the sliding body. A body starting from rest will reach again the same height. This has a remarkable static analogy, the hydrostatic balance and the principle that "water seeks its own level." But even without the concept of work used in medieval statics, there was something "conserved" here. A body's speed at the base of the plane depends only on the height from which the body begins. The time to reach any point along the plane also depends only on the initial height, not the inclination of the plane. Two bodies rolling at constant speed reaching the edge of a table will have different ranges depending on their horizontal motion but will hit the ground at the same instant. Thus, he reasoned, one and only one of the motions is accelerated, that along the vertical. The other continues as it had before the fall began. And further, the accelerated motion is produced by a force acting on a body independent of its mass. We see the serendipity of working with weights since that depends only on the acceleration of gravity and the mass.

So for Galileo, freefall is *accelerated* motion. The magnitude of the speed, not just the distance covered, changes in time. This is a comparison between distinct quantities. The speed increases directly proportional to the interval of time. This was the odd-number rule, that the proportional distances, s_n, covered are in ratios of odd lumbers for successive equal intervals of time (e.g., in intervals of 1,3,5, etc. for successive equal time intervals), $s_{n+1} : s_n = (2n + 1) : (2n - 1)$, for the nth interval. The weight of the body, and by inference its shape, don't affect the process. Again, this reduces the problem to proportions and geometry. The speed during an interval of time, T changes from, say, zero to some V. Therefore the average speed in this interval increases as $T/2$. Now since the distance covered in an interval T is given by the average speed, we find two results: that the proportion of speeds in two intervals is $V_2/V_2 = T_2/T_1$, and that the distances traversed if the acceleration is constant is $L_2/L_1 = (T_2/T_1)^2$ that is the compound ratio $L_2/L_1 = (V_2/V_1)(T_2/T_1)$.[3] It is then a relatively short step to the final form of the law, that $L = \frac{1}{2}gT^2$ for a constant acceleration of gravity g. Notice this actually requires including the *dimensions* of acceleration, (length)/(interval of

DEL GALILEO. 129

E per vn breue eſempio di queſto che dico diſegnai già la figura di vn' oſſo allungato ſolamente tre volte, & ingroſſato con tal proporzione, che poteſſe nel ſuo animale grande far l' uffizio proporzio-

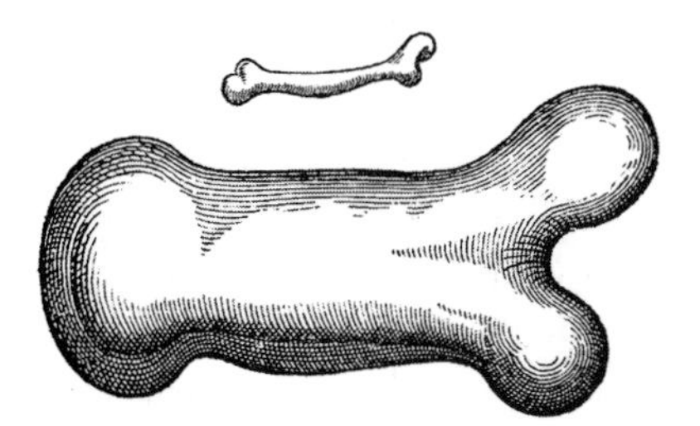

nato à quel dell' oſſo minore nell' animal più piccolo, e le figure ſon queſte: doue vedete ſproporzionata figura, che diuiene quella dell' oſſo ingrandito. Dal che è manifeſto, che chi voleſſe mantener in vn vaſtiſſimo Gigante le proporzioni, che hanno le membra in vn huomo ordinario, biſognerebbe ò trouar materia molto più dura, e reſiſtente per formarne l'oſſa, ò vero ammettere, che la robuſtezza ſua fuſſe à proporzione aſſai più fiacca, che ne gli huomini di ſtatura mediocre; altrimente creſcendogli à ſmiſurata altezza ſi vedrebbono dal proprio peſo opprimere, e cadere. Doue che all' incontro ſi vede nel diminuire i corpi non ſi diminuir con la medeſima proporzione le forze, anzi ne i minori creſcer la gagliardia con proporzion maggiore. Onde io credo che vn piccolo cane porterebbe addoſſo due, ò trè cani eguali à ſe, mà non penſo già che vn cauallo portaſſe ne anco vn ſolo cauallo à ſe ſteſſo eguale.

Simp. *Mà ſe così è, grand' occaſione mi danno da dubitare le moli immenſe, che vediamo ne i peſci, che tal Balena, per quanto intendo, ſarà grande per dieci Elefanti, e pur ſi ſoſtengono.*

Salu. *Il voſtro dubbio S. Sim. mi fà accorgere d'una condizione da me non auuertita prima, potente eſſa ancora à far che Giganti,*

R & altri

Figure 3.4: An illustration from Galileo's *Two New Sciences* showing the scaling of a bone as an illustration of the application of statics and breaking tension to the size of animals. Image copyright History of Science Collections, University of Oklahoma Libraries.

time)^2, analogous to the *difform difform* of the Mertonians. Most important is that the law is not a law for force nor does it help yet in defining forces. It's purely kinematic with no pointer to the origin of g.

In his youth, sometime while in Pisa before 1590, Galileo realized that the oscillation period of a pendulum is independent of both the mass of the bob and its initial displacement from the vertical. All that matters is the length of the rod or chord on which the mass is suspended. This *isochronism* made a powerful impression on him and we know it occupied his attention for decades. The observation that regardless of the weight the motion is identical for equally long suspensions was paradoxical within the received physical laws. In apparent contrast to freefall, and more similar to the inclined plane, pendular motion is constrained. The body moves along an arc whose length is fixed. Since the only driving motion is its weight, its acceleration should have depended on the weight of the bob. It didn't. For anyone whose learning had been based on Scholastic principles this was completely counter-intuitive in the same way we encountered for a freely falling body. But there are two similarities with the inclined plane. First, once released the body moves through the minimum and returns to rest the same height from which it started. Its speed increases steadily through the lowest point and then decreases as it climbs. If given an initial impulse, the body will rise farther, precisely as for an inclined plane, and when on the horizontal (which for the pendulum is instantaneous) the speed will be constant. One can imagine that the pendulum breaks at this point, a question that isn't in Galileo's treatises. The period of the pendulum therefore depends only on the length of the constraint, independent of the initial displacement. This is the hardest point to see without the notion of inertia.[4]

Galileo's Statics

Throughout antiquity, and extending into the Middle Ages, there were two, often non-intersecting, discussions of force. One was dynamical, the problem of motion. Because after Aristotle the cause of movement was connected with theological ideas, it was the fodder for a vast number of commentaries and doctrinal disputes. The other was more subtle, statics. Although this was certainly the more quantitatively mature even before the late antique period, it concerned equilibrium and more mundane—indeed exceedingly practical—problems, it was far less clearly appreciated as a general application of the idea of forces. Precisely because it dealt with balance it was thought, at first, to be a separate problem entirely. The moment when these two strands merged can be pinpointed precisely, the publication of Galileo's dialog *Discourses and Mathematical Demonstrations Concerning Two New Sciences* in 1638. The two sciences of the title were statics and dynamics.

Recall that Archimedes dealt, in broad terms, with two areas: hydrostatics and simple machines. In these he was able to ignore any precise definition of *force* while at the same time making use of weight but he ignored the structure of the machines. How did Galileo consider the problem? First, he began with a story about the breaking of a large marble pillar laid on two supporting columns that was broken when a third was incautiously added in the middle. This introduced a long examination of the strength of materials and, in the second day of the dialog, the application of static load, torque, and scaling to the general treatment of strength of materials and static equilibrium. For the first time, a general force, not just weight, was used to describe the breaking of a body.

Strain is the distension, or geometric distortion, of a body. If it's only one dimensional, a string for instance, this involves only in its contraction or lengthening. If, on the other hand, the object has a finite extension, a beam for example, this involves a distortion in at least two dimensions and requires the concept of the lever. It's odd that even though Galileo discussed the bending moment of a beam, he never proposed any general law for how the form of the body affects its reaction, nor did he realize the principle that Newton would about 50 years later would call the third law of motion, that a body reacts with the same force that an external agent exerts. Neither did he realize that if a body is distorted, it must also be sheared and this will cause a change in his estimate of the point of failure. Although resistance is defined as the ability of an object to withstand a load—as we will see later, actually a statement of Newton's third law of motion: a force is mutual, that which is acted on resists the force—it isn't extended to compare magnitudes.

Let's examine this example a bit more closely. Galileo analyzed the breaking of a loaded column by imagining a weight attached at its bottom and hung vertically. This produces a symmetric lengthening of the column, although it might not be visible, and ultimately will beak it if the weight is too strong. So the stress on the column is uniformly distributed over the area. If there are more attachment points for the material of the column, in other words if the cross section is larger, then the column will have a greater resistance to breaking. OK, so far. Now extend this to the idea of a beam stuck in a wall. In this case, assume the beam is bending under its own weight. Using the idea that the wall–beam combination acts like a heavy lever, in which the weight of the lever isn't negligible, the point of attachment to the wall below the beam is the fulcrum. In this case, the stress is concentrated there. The resisting force along the cross section of the beam is uniform and the moment arm is now the half-thickness of the beam. The experience from the stretched column then says the stress—resistance force—is uniformly distributed across

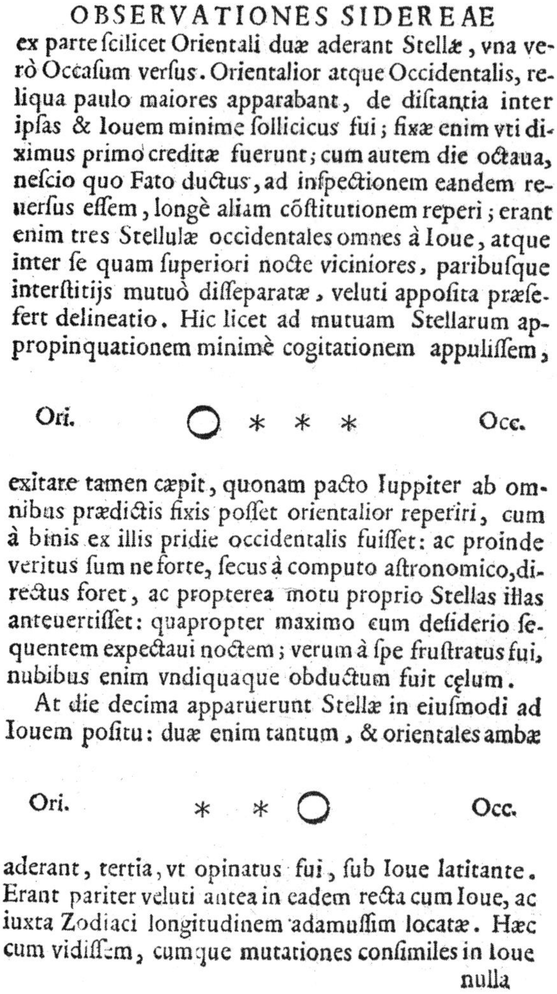

OBSERVATIONES SIDEREAE

ex parte ſcilicet Orientali duæ aderant Stellæ, vna verò Occaſum verſus. Orientalior atque Occidentalis, reliqua paulo maiores apparabant, de diſtantia inter ipſas & Iouem minime ſollicicus fui; fixæ enim vti diximus primò creditæ fuerunt; cum autem die octaua, neſcio quo Fato ductus, ad inſpectionem eandem reuerſus eſſem, longè aliam cōſtitutionem reperi; erant enim tres Stellulæ occidentales omnes à Ioue, atque inter ſe quam ſuperiori nocte viciniores, paribuſque interſtitijs mutuò diſſeparatæ, veluti appoſita præſefert delineatio. Hic licet ad mutuam Stellarum appropinquationem minimè cogitationem appuliſſem,

Ori. O * * * Occ.

exitare tamen cæpit, quonam pacto Iuppiter ab omnibus prædictis fixis poſſet orientalior reperiri, cum à binis ex illis pridie occidentalis fuiſſet: ac proinde veritus ſum ne forte, ſecus à computo aſtronomico, directus foret, ac propterea motu proprio Stellas illas anteuertiſſet: quapropter maximo cum deſiderio ſequentem expectaui noctem; verum à ſpe fruſtratus fui, nubibus enim vndiquaque obductum fuit cęlum.

At die decima apparuerunt Stellæ in eiuſmodi ad Iouem poſitu: duæ enim tantum, & orientales ambæ

Ori. * * O Occ.

aderant, tertia, vt opinatus fui, ſub Ioue latitante. Erant pariter veluti antea in eadem recta cum Ioue, ac iuxta Zodiaci longitudinem adamuſſim locatæ. Hæc cum vidiſſem, cumque mutationes conſimiles in Ioue nulla

Figure 3.5: The discovery of the presence and motion of the satellites of Jupiter from Galileo's *Sidereus Nuncius*. This was the first demonstration of bound orbital motion around another planet and showed the reasonableness of the Copernican construction for a heliocentric planetary system. Image copyright History of Science Collections, University of Oklahoma Libraries.

the cross section. Although the result Galileo derived is quantitatively wrong in absolute units, the basic result *is qualitatively correct if stated as a proportion*: the resistance to breaking of a beam of any length depends on the width of the beam in the direction parallel to that of the load. Thus for two beams of widths a_1 and a_2, the resistance scales as $R_2/R_1 = a_2/a_1$. It's an easy demonstration, take a flexible meter stick and allowing enough of the beam to extend over the edge of a table, first place the broad side on the table and then perpendicular. You'll easily see the difference in the deflection of the end. This is the principle that allowed for the next step of the *Two New Sciences* where we see one of the most novel applications of physics of Renaissance natural science. Galileo tackled a knotty problem, one that would surely have delighted Aristotle himself and that eventually inspired the Victorians (especially after Darwin had introduced the concept of evolution by natural selection): what is the optimal design for an animal given its size and mode of locomotion? Galileo didn't state it quite this way, but you can see where he is going. Because the material itself has weight, there is a breaking stress that depends on the area and volume of the body, in this case imagine just one bone. A simple uniform geometric scaling, where the dimensions are each identically multiplied, m will increase the mass by a factor m^3 while the area will only increase as m^2. Thus, without some change in the conformation of the body—its shape—the animal will be crushed under its own weight before it can take a single step.[5] With this beautifully simple deduction Galileo founded theoretical biomechanics.[6]

Celestial Phenomena and Terrestrial Physics

In 1609, Galileo was mainly engaged in a wide range of mechanical studies while also lecturing on natural philosophy at Padova and taking private students. Much of his work to that time had been circulated but unpublished, although he had published descriptions of a number of his mechanical inventions. It was the year news reached Venice of the Dutch invention of a spyglass, an optical tube that magnified distant objects. It isn't important that Galileo did not *invent* the telescope. He invented observational celestial physics. The "cannocchiale" became an instrument of discovery and he made extraordinary use of it. The results were announced in the *Sidereus Nuncius* (the "Starry Messenger") in 1610. He observed the surface of the Moon and found it to resemble the Earth, showing mountain ranges and valleys, dark patches that resembled seas and bright spots that he explained as individual peaks caught in sunlight. These discoveries challenged the fundamental principles of Aristotelian physics. As long as the Moon, the closest resolved celestial body, could be considered somehow different from terrestrial stuff it was alright for it to circle the immovable Earth at the center of the world. After Galileo this wasn't permissible. The Moon is like an Earth. It must be heavy and must partake of the same natural movement as all heavy things. There was no question, though, that it orbits the Earth and after Tycho there were no longer spheres to impede its descent.

This contradiction couldn't be ignored by his contemporaries. The theme of the dynamical connections in the Earth–Moon system would continue to be a major issue for physics. It is important that the parallactic distance of the Moon from the Earth was already well known that Galileo could put physical dimensions on what he was seeing on the lunar surface. His geometric construction of how to measure the height of a lunar mountain was the first example of what we would call "remote sensing." But more important still was that now the scale of the body could be understood, its physical size relative to the Earth was already appreciated but not how much stuff it contained. Lacking, however, any further explanation for the dynamics, any more than Copernicus himself, Galileo remained content to explore the extraterrestrial landscape and describe its structure without entering into the dynamics. At least not yet.

In a further extension of the concept of space, Galileo's resolution of the nebulous patches in the Milky Way into an uncountable number of stars gave the first signal of a depth to the stellar realm. Increasing aperture and magnification simply revealed more, fainter stars in ever direction. Further, they were clumped, not uniform and continuous as they had appeared to the unaided eye. If the Copernican picture were right, he realized, these observations implied there is no limit to the extent of the stellar population, that each is a Sun in its own right (as Giordano Bruno had held).

But the most important result for our history was his chance observation of four faint (he called them "small") stars accompanying Jupiter. With an eye toward future employment at the Tuscan court, he named these the Medicean Stars. They not only moved with the planet, they circled it periodically. For the first time since the beginning of astronomy, a motion was observed that could not simply be transformed away by a change of coordinate. The motion was not merely projected against the stars, it was a motion in space around a center that is *not* the Earth. Galileo completely understood the revolutionary nature of his discovery. While the Copernican system implied a new physics, his observations required it. Again there was a challenge to the inherited dynamics.

There followed a steady stream of new telescopic discoveries. In 1612, in his *Letters on Sunspots*, Galileo announced the observation of irregular, ephemeral dark markings—sunspots—that were sited on the surface of the Sun and with which he determined the Sun's rotational axis and period of rotation. This furthered fueled the challenge to the ancient mechanics: now even the Sun, although luminous, needed to be included in the scheme of a terrestrial physics. The solar rotation also showed that a body could possess more than one motion proper to itself, a further argument in favor of the rotation of the Earth. In the same book, he also announce that Venus shows phases like those of the Moon. Understanding that the lunar phases are due to phase dependent solar illumination around an orbit, he showed that the variation in the angular diameter of the planet and its changing illumination demonstrates its orbit is centered on the Sun, not the Earth. Along with the Jovian satellites, there was now no question regarding the heliocentric system: it was correct.[7]

The fruit of Galileo's observations and thinking about their implications was published in 1632, *The Dialog Concerning the Two Chief World Systems*. Along with its orbital motion, Copernicus asserted that the daily motion of the Sun, planets, and stars together require the rotation of the Earth. The seasonal variation of the solar track he explained by inclining the rotation axis to the orbital pole. A further consequence of the motion of the Earth was its multiple motions. It both moves in its orbit and rotates daily. Neither seems to have a physical consequence. That is, in Scholastic arguments, with the exception of Jean Buridan and Nicholas of Cusa (1401–1464), the motion of the frame is distinct from the bodies within it. A falling body, in this dynamics, is independent when in motion and if the frame—in this case, the Earth—were to shift by rotation the body would fall behind the direction of motion. And Ptolemy had persuasively argued that the air would not move with the Earth and therefore would always produce a wind issuing from the direction of rotation. Since these consequences were both obviously in contradiction to everyday experience, common sense implied the stationarity of the Earth in all senses. Galileo argued, instead, that motion is shared by all bodies with the frame in which they are comoving and that no measurement made *within* that frame can dynamically sense its motion. The simplest example is to imagine yourself in an airplane moving at constant speed. If you drop a cup, it falls at your feet even if the plane is moving at hundreds of meters per second. This is the first example of an invariance principle in physics and is still called "Galilean invariance." With it motions in a moving system can be separated into those proper to the frame itself and those within the system. Even if, as in freefall, the body accelerates the rotation of the Earth is slow enough that the motion is vertical for a small enough distance from the surface, such as the example.

Despite these discoveries, Galileo didn't concern himself with an explanation of the heliocentric motion of the planets and the motion of the Jovian moons. He did attempt to find an Archimedian sort of explanation for the tides which came directly from his advocacy of the Copernican system applying a sort of inertial principle. Since the motion of the Earth–Moon system was not uniform relative to the rotation of the Earth the oceans would execute a daily sloshing motion due to a torque. This couldn't succeed as an explanation of the phenomenon but it is important that Galileo attempted to apply the principles of force and inertia to a cosmic-scale problem. In this as in everything else he did Galileo's approach was distinctly experimental. He was able to use inertia to demolish arguments against the motion of the Earth and apply it equally to terrestrial and cosmic motions. He made mechanics an empirical physical science. But on the bigger scheme he had less to say and was not able to appreciate Kepler's approach to physical astronomy and the question of the origin of the force that maintained the cosmos remained an open question.

In summary, the result of several centuries of Scholastic discourse was to establish the equivalence of change with the action of a force. Additional effects, additional forces, could introduce changes in the changes, the difference between "uniform difform" and "difform difform." Regardless of what it is, for the Scholastics, as it was for Aristotle and the ancients, the cause of *any* change required some continual action to maintain constant rate of change. For them, nothing was conserved in motion. A natural motion is in the essence of the thing but results from displacement. It is the only form of movement that requires no driver other than a final cause within the thing itself. Once the force overcomes some threshold, unspecified but necessary, the motion commences and continues only as long as the superior driver is maintained. Galileo's conception was fundamentally different: the motion, once started, continues, requiring nothing to maintain it in the absence of resistance. This last requirement is the same as saying: if there is no impeding action, or force, a body maintains its constant state of motion. *This is inertia*. The question of weight, or mass, doesn't enter here because a massless body simply doesn't exist. Instead, extension does make a difference. The state of motion is different for sliding or rolling motion, for example, but both have their inertia and will conserve that state similarly. The same is then true for the celestial motions, these do not require a separate explanation from terrestrial motions. Nor can they require anything different, since having telescopically discovered the similarities with earthly things these bodies should behave similarly.

Galileo's other great contribution was to introduce quantitative as well as qualitative elements into the study of nature. When treating the rate of change, the Scholastics could arbitrarily create numbers. Galileo, instead, made actual measurements and had real data against which to test his ideas. The way was now open for considering a new approach to mechanics in particular and natural philosophy in general. If the heavens and Earth are really a continuous whole, then any experience here applies to the universe as a whole. More importantly, Galileo had set the research program for the centuries to come: the universalization of the physical laws. Priority disputes aside, what matters isn't that the effects were new, at times others observed the phenomena before or simultaneously. The difference lies in Galileo's use of the empirical results, his singular ability to extract general physical principles from the experiments and observations and to see their broadest application in natural philosophy. Before him, the universe had been divided. After Galileo, it would never be again.

Pressure, Temperature, the Atmosphere, and the Force of the Void

Galileo's example inspired a completely new approach and his followers wasted little time in extending mechanics to practical problems. Castelli and Borelli applied force concepts to moving fluids, especially canals and the control of rivers (a perpetual problem in European cities, especially in Italy, during the Renaissance). Galileo's studies of strength of materials led to works on theoretical architecture and the design of machines and ballistics. I'll mention here just one of the areas that grew out of this new understanding of forces, the study of atmospheric phenomena and the behavior of different states of matter.

The expansion of heated bodies was one of the discoveries that led to Galileo's invention of a practical device, the thermometer, that consisted of a glass bulk containing liquid into which an inverted tube capped by another bulb was placed. When the fluid was heated it expanded up the tube. Its design inspired Evangeliste Torricelli (1608–1647), one of Galileo's last assistants and pupils, to use a similar device to study the atmosphere. He replaced the close system with an open pan of mercury, to increase the inertia of the fluid, and found that the column reached a fixed level. He reasoned that the weight of the atmosphere above the pan produced the pressure that supported the fluid against the vacuum at the top of the column. This is a *barometer*, the first practical device for measuring gas pressure. He announced the invention to a correspondent at Rome in 1644. This, by extension, also connected with a problem that had bothered Galileo in understanding the cause of the mechanical behavior of materials, the role of "nothingness." Emptiness was impossible for Aristotelian physics. Without resistance a force would produce an infinite velocity that, obviously, was impossible; Aristotle had removed the void by a *reductio ad absurdum* argument. But the atomists had not only accepted the idea but required it. In the first day of the *Two New Sciences*, Galileo began his discussion of the strength of beams with:

> I shall first speak of the void showing, by clear experiences, the nature and extent of its force. To begin with, we may see whenever we wish that two slabs of marble, metal, or glass exquisitely smoothed, cleaned, and polished and placed one on the other, move effortlessly by sliding, a sure argument that nothing gluey joins them. But if we want to separate them parallel, we meet with resistance, the upper slab in being raise draws the other with it and holds it permanently even if it is large and heavy, This clearly shows Nature's horror at being forced to allow, even for a brief time, the void space that must exist between the slabs before the running together of parts of the ambient air shall occupy and fill that space.

He then explained how to measure this force, using a cylinder in which there is placed a weighted piston using a bucket that can be filled with sand. Air is evacuated from the interior by filling it with water and then, after the air is out, inverting the sealed tube and loading the piston. The force counteracting the load is then the "force" of the vacuum. This is an ingenious argument. Galileo uses it to explain the breaking strength of a beam of glass or stone, that this force extended over the area of the interaction produces a stress that retains the integrity of the material. For a continuous medium with weight, the atmosphere for example, this made sense because a reduction of the density in one place produced an acceleration toward the deficit and with such arguments it was possible to produce and explain the barometer.

The theatrically brilliant experiment of Otto von Guericke (1602–1686) at the town of Magdeburg illustrated this point, again with the inversion of the meaning of the void. He evacuated a large sphere, about one meter in diameter, split into two hemispheres and tied to two opposing teams of horses. The horses failed to separate them. It's important to note that this was, for him, a demonstration of "nothingness." Returning to Aristotle, this showed that space (because clearly the spheres occupied space) was not corporeal but merely dimensioned. For Von Guericke, this had great theological significance, as he wrote in his *Nova Experimentum* or *New Magdeburg Experiments (so called) on Void Space* (1672):

> the universal vessel or container of all things, which must not be conceived according to quantity, or length, width, or depth, nor is it to be considered with respect to any substance . . . But it is to be considered only in so far as it is infinite and the container of all things, in which all things exist, live, and are moved and which supports no variation, alteration, or mutation.

This was almost a quarter century after Galileo and now the concept of "force of the void" had started to become identified with "space." The force involved a medium between between things, whether they were within and out of air, not in or out of vacuum.[8]

Pierre Gassendi (1592–1655) continued this line of thought in an argument that influenced many of the subsequent discussions of the existence of the void:

> Space and time must be considered real things, or actual entities, for although they are not the same sort of things as substance or accident are commonly considered, they actually exist and do not depend on the mind like a chimera since space endures steadfastly and time flows on whether the mind thinks of them or not.

If we say that things exist in space and time, both must be separate from matter. Therefore motion exists in both and is a property of the matter within the geometry of the world. We need only a concept of time to complete this, but it is independent of the substance. The void is just that, empty space imbedded in time, and there is no force to the vacuum. An atomist, Gassendi's view of material was inherited from the Greek and Roman writers, notably Democritus, Epicurus, and Lucretius, but with significant modifications:

> The idea that atoms are eternal and uncreated is to be rejected and also the idea that they are infinite in number and occur in any sort of shape; once this is done, it will be admitted that atoms are the primary form of matter, which God created finite from the beginning, which he formed into the visible world, which finally he ordained and permitted to undergo transformations out of which, in short, all the bodies which exist in the universe are composed.

The final phase of this exploration of "nothingness" included the air pump demonstrations of Robert Boyle (1627–1691) and the studies of atmospheric pressure and fluids by Blaise Pascal (1623–1662). Pascal was the first to attempt a measurement of the variation of the weight of the atmosphere with height, soon after the announcement of Torricelli's invention of the barometer reached France. In 1647–1648, he used a barometer to measure the change in the pressure going from the base to the summit of the Puy-du-Dome, near his home in Clermont-Ferand in central France. He also studied hydrostatic pressure, showing that the effect of immersion produced a reaction that is always normal to the surface of the body regardless of its shape, a fundamental extension of Archimedian principles, the results appearing posthumously. By the middle of the seventeenth century the application of mechanical laws to the dynamics of both particles and continuous media had been started.

CARTESIAN MECHANICS

With Rene Descartes (1596–1650), we encounter a philosopher of Aristotelian proportions whose system encompassed the entire universe and all of its diverse phenomena. Unlike Leonardo, for whom nature was the ultimate referent, Descartes' prospectus centers on the observer and the basis of perception. In his hands, geometry became a new, potent analytical tool. Descartes' introduction of an algebraicized geometry, quantifying and extending the techniques pioneered by the Calculators and their contemporaries, reduced Euclid to operations on symbols and provided a simpler means for producing quantitative results from such arguments. Descartes set out the full scope of his mechanical vision in *le Monde* (The World) and, later, in the *Principles of Natural Philosophy*. Its range is enormous, from perception and reason to motion and ultimately to the system of the world. It would soon serve as both a foil and model for Newton in his deliberately more focused work of the same name, with a vastness that inspired a generation to adopt the *mechanical philosophy*. Even Galileo had not speculated on so vast a range of topics in his dialogs, preferring instead to distinguish between the physical and metaphysical. Descartes did not recognize such distinctions.

The central concept of Cartesian mechanics was, as with Galileo, inertia. But Descartes asserted that God, not the material thing, is the first cause of movement who always preserves an equal amount of movement in the world. This is very different from Galileo for whom inertia is an intrinsic property of matter once set into motion, not a global property parsed out to the different parts of creation. Descartes arrived at this notion by experimenting with collisions, both real and thought experiments. Imagine two balls of equal weight, one at rest and the other moving. In a collision, the mover comes to rest and the other picks up its motion. If the weights are unequal, there is a redistribution of the impulse between them with the concomitant division of the quantity of motion. Thus a state of motion can also produce one of rest and a collision between unequal bodies can even reverse the direction of motion for one of them. This is still not a force principle, as such, because impact can be approximated as instantaneous. From these observations he deduced three laws of motion. The first states that "each thing as far as in it lies, continues in the same state; and that which is once moved always continues so to move." This is the assertion of inertia but with the states of rest and motion distinguished. Note the wording carefully, when we discuss Newton you will see how different two versions can be of the same law with just a minor change in the wording. The second law states that "all motion is of itself in a straight line; and thus things which move in a circle always tend to recede from the center of the circle that they describe." Now we see what Descartes means by force. It is *centrifugal*, seeking always to move from the center of curvature. But something must act to maintain the curvilinear motion; that is the force. The third, and final, law states that "a body that comes in contact with another stronger than itself loses nothing of its movement; if it meets one less strong, it loses as much as it passes over to that body." This came from thinking about collisions. He went further in asserting the equivalence of rest and motion, that these are states of a body and that the action required to accelerate a body (change its state) is the same for each. In both cases, he was describing rectilinear motions, only changes in speed. Descartes never asserted that these two different states are identical because rest is *relative*, there is no concept here of a frame of motion. That was left to Galileo. In a collision, the effect is one of reflection, we would now say this is in the "limit of infinite mass ratio" that a small body reflects, without apparently moving the larger mass. For Descartes a collision was, then, an all-or-nothing affair.

Figure 3.6: Rene Descartes. Image copyright History of Science Collections, University of Oklahoma Libraries.

For two dimensional motion, Descartes concentrated on curves. He stated that such motion requires constraints without using the Aristotelian categories (*natural*, *accidental*, or *violent*) and then passes to what he calls the Second Law of Nature, that " a moving body tends to continue its motion in a straight line." As illustration, he uses a stone held in a sling, arguing that from the distance traversed along a plane tangent to the curve that the motion is accelerated and therefore requires an action, which is provided by the sling and the hand. The sling and hand (material and final causes, if we were to employ Hellenistic concepts) are the movement and mover. That when released the ball continues in a straight line is because the constraints instantaneously vanish, analogous reasoning then can be transferred from collisions to allow ignoring the moment of escape. Later in the work, he presents it without using the term mass, that a body meeting a stronger one than itself loses nothing of its motion while if it encounters a weaker one, it gives up its motion. The problem is stated simply: how do we find the force required to maintain a body in motion along a curved path? In particular, when the path is a circle? Descartes realized the motion of a body is accelerated in some way when it moves in a circle, but he dismissed a change in direction as something that requires a constraining force. Think of that tethered ball. If we ask what the speed is, then along an arc of a circle—provided the length of the string remains unchanged—the time to cover equal portions of the arc is the same (constant circular speed). Yet there is a tension on the string. That comes from resisting the tendency of the ball to escape by moving away from the center.

Although he began as a Cartesian enthusiast—indeed he was Descartes' student—Christian Huygens (1629–1695) provided, through the repudiation of some basic Cartesian principles, the first quantitative law for force in circular motion. In 1657, Huygens dealt with this by requiring the motion to be an outward

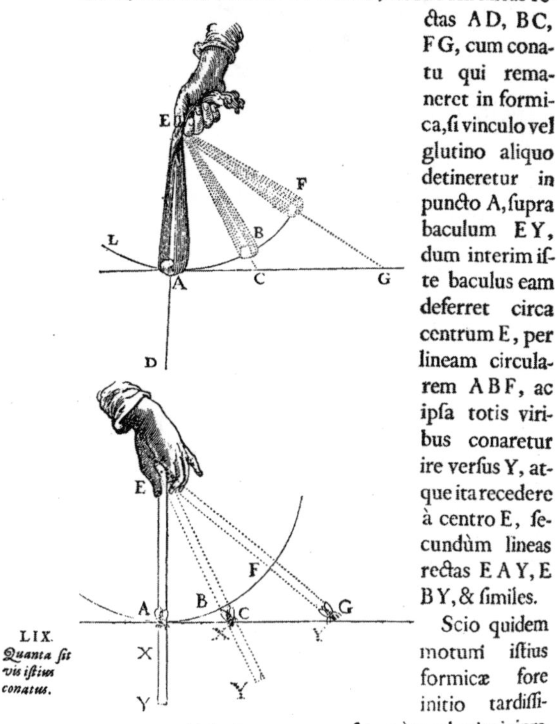
100 PRINCIPIORUM PHILOSOPHIÆ

idem lapis, actus in funda ſecundùm lineam circularem ABF, recedere conatur à centro E, ſecundùm lineas rectas AD, BC, FG, cum conatu qui remaneret in formica, ſi vinculo vel glutino aliquo detineretur in puncto A, ſupra baculum EY, dum interim iſte baculus eam deferret circa centrum E, per lineam circularem ABF, ac ipſa totis viribus conaretur ire verſus Y, atque ita recedere à centro E, ſecundùm lineas rectas EAY, EBY, & ſimiles.

LIX. *Quanta ſit vis iſtius conatus.*

Scio quidem motum iſtius formicæ fore initio tardiſſimum, atque ideò ejus conatum, ſi tantùm ad principium motus

Figure 3.7: Figure from Descartes' *Principia* showing his explanation for centrifugal force. A body tends to recede from the center of a curve, the sling maintains the motion around the curve against this. While this is true in the moving frame, Newton demonstrated that the unconstrained motion is really along the instantaneous tangent to the curve in a straight line, not radially outward. Image copyright History of Science Collections, University of Oklahoma Libraries.

tendency of the body in contrast to the tension of the sling. Thus was formed the concept of *centrifugal* force, which would become a central concern for Newton. Huygens took the change in direction, not simply the speed, to be an attribute of accelerated motion and, since this is outward, took the acceleration to be similarly directed. The rate at which a body recedes from the curve along the tangent plane clearly accelerates, the projected velocity varying in magnitude to reflect the constrained nature of the motion, the *conjunctus*. The acceleration depends on the tangential speed v and the radius of the arc, r, varying as v^2/r. You'll notice I've called this an "acceleration," anticipating later developments; the tendency is a vague thing since the definition of the mass is left unresolved in Descartes and Huygens. But there is a more important feature. This is an *outward* tendency from the point of view of the observer moving along the arc. The body, now freely moving along the tangent plane, *seems* to be receding at increasing rate while, in fact, it is moving at constant speed and in a constant direction.

In Cartesian celestial mechanics the universe is container filled with a complex dynamical fluid, an ether, but not the sort imagined by the Hellenistic philosophers. This one is not only in constant motion but filled with vortices. It was his explanation for the attraction that binds the planets to the Sun and the Moon to the Earth. It's justification comes from a familiar phenomenon. Take a cup of water and put tea leaves into it. Stir it so the tea moves to the sides of the container. So far, no problem, this is a centrifugal force that Descartes had postulated is an essential property of circulating matter. But if you stop the stirring, the leaves will move to the center and bottom of the cup. The fluid at the walls is more slowly than in the center of the vortex. It seems to be an attraction that draws the leaves toward the center, just as gravity does for heavy bodies. To Descartes this was a powerful analogy. As Giordano Bruno had before him, Descartes was seized with the image of each star being a sun at the center of a vast sea of cosmic vortices each supporting possible planetary systems. The inspiration of Galileo's astronomical discoveries and Kepler's laws of planetary motion combined with a dynamical driver completed the construction. Cometary paths could be irregular because they pass from one vortex to another. The Moon is drawn to the Earth by its local vortex while the pair orbit the Sun for similar reasons. Although Descartes made no attempt to check whether this picture is really consistent with Keplerian motion and the harmonic law, it was enough to provide a compelling physical construction and a hypothesis for the origin of gravitation without natural motion.

On the other hand, the essentially Aristotelian concept of a force still applies, this is a medium that is actively driving the motions so the inertia of circular motion that Galileo had introduced is missing. Although not a magnetic driver, the vortices bear a striking resemblance to Kepler's dynamical model. Furthermore, the nature of this medium is unexplained, whether it has gravity itself or only produces it. This concept of a space-filling vortical plenum remained an attractive explanation for nearly 150 years. It satisfied almost all explanatory requirements even if it lacked quantitative details. It spanned the gap between Scholastic mechanics and the new physics. It impeded, however, the acceptance of the existence of a vacuum, since if the universe is completely filled only relative augmentations or diminutions are possible, and in adopting a continuum denied the existence of atoms. Hence the system explained well the motion of the planets within a heliocentric system, linked cosmology, celestial mechanics, and rather general laws of motion even as it left huge conceptual and foundational holes.

ROBERT HOOKE: THE FORCE LAW FOR SPRINGS

Among his numerous optical, naturalist, linguistic, and astronomical studies, Robert Hooke (1635–1703) made a vital contribution to mechanics with a commonplace discovery, his elastic law announced in his *Cutlerian Lectures* published in 1679: the extension of a spring, ΔL, depends linearly on the weight attached to its end and the material of which the spring is made, weight $=$ constant $\times$ ΔL If we replace the weight by any other action, while requiring that the spring that it can only stretch in response to an applied force, we can replace the weight by any force. Then we arrive at Hooke's statement that *strain, the deformation of an elastic body, is directly proportional to stress, the impressed force.*

Consider for a moment the number of concepts underlying this very simple result. Before the promulgation of Hooke's law, the measure of weight was essentially along Archimedian lines with the level or steelyard. Although using unequal arms, the principle is unchanged from the early mechanical treatises. The two forces are identical since they're both weights and therefore gravity alone is involved. You don't need to consider either the material (unless the level or arm distorts) or anything else about the constitution of the material. But *if* you use a simple elastic law for the measure of the force, you're actually contrasting two different forces—elasticity and weight—and saying that one is the reaction to the other. This is *not* the same as a beam balance and the difference is very important. We are actually making a generalized statement, that all forces are equivalent, and then developing a technique based on this. Notice also that the spring balance uses the reaction—extension—of the spring and not its acceleration. But if we release the end, the spring contracts, or in the language we have already used "changes its state of motion," by accelerating. This is easily seen. It also doesn't matter what form we choose for the spring, helical or linear, or whether the medium is simply elastic (so we have just one dimensional deformation) or helical (where we twist and extend the spring simultaneously). Elastic deformation actually affords a connection with the discussion in Day 1 of the *Two New Sciences*, when Galileo talks about the breaking of a beam, because we now have a model to complement his more qualitative arguments from geometry. In fact, Galileo had already assumed something like a spring law for building up the structure of an elastic body (the fibers of the wood act as individual, extendible elements of the material) but with Hooke's law we have a direct, verifiable, *quantifiable* principle.

NOTES

1. Although contemporaries thought this preface, which presented an even stronger qualification of the work as a hypothesis, was written by Copernicus himself, Kepler later showed that it was actually a separate, clandestine addition to the work a the time of its publication.

2. For instance, two of Galileo's earliest productions were works on the hydrostatic balance, *bilancetta*, and a device for computing ballistic range and flight, the *military compass*. He also served as a consultant in practical, engineering matters to the Venetian senate much as Archimedes had to the ruling class of Syracuse.

3. Stillman Drake, the principal biographer of Galileo of the last century, made an excellent point about the use of algebraic notations in understanding the Galilean argument. Any ratio was held, following Euclid, *Elements* Book V and Eudoxus, to be a comparison of physically similar (commensurate) quantities. To belabor the point for a moment, this means you can compare distances and times separately but not together because they are not the same thing. While now we employ the algebraic notion of units, as we have seen motions are decomposed into space and time separately and these were not on the same footing in the seventeenth century. Thus, to arrive at the quadratic relation for the time dependence of displacement, and then to extend this to a general dynamical law for acceleration, requires using a compound ratio of two separate quantities.

4. We consider the velocity of the body at any moment to depend only on the time. The distance covered along an arc, s, is the angle through which it moves, $\Delta\theta$ times the radius of the circle, R. Since the weight is both falling and moving horizontally, and that second motion is only due to the constraint, the vertical distance of the fall should depend on the square of the time. We note that, for a small angular displacement, the bob is raised through a vertical distance $R\Delta\theta$. Like a body on an inclined plane, the positional gravity will be larger at the beginning than at the end of the motion.

5. In the twentieth century, it might have been more interesting if science fiction writers had read the text of *Two New Sciences* more carefully. Godzilla is a scaled up version of a dinosaur. The Amazing Colossal Man and the Fifty Foot Woman are examples of how little attention was paid to such consequences (a lovely essay from 1928 by J. B. S. Haldane, "On Being the Right Size," is a notable exception). Yet Galileo himself points out the problem when discussing marine creatures, since he shows that on land they will be unable to function, being without the hydrostatic support provided by the water in which they live.

6. You may object that Galileo was working almost 150 years after Leonardo da Vinci had made similar proposals. The difference, as I've emphasized, is that Leonardo never prepared anything systematic, or even synthetic, on his physical investigations. He'd thought about publishing them but never got around to it, preferring always to be wandering in uncharted intellectual territory. His reputation by the close of the sixteenth century was

as a supremely gifted artist, if not eccentric, inventor and applied mechanician. Many of Galileo's derivations and ideas were along the same lines but that is an indication of how convergence occurs in scientific problems rather than an intellectual indebtedness.

7. It is, however, important to note that this observation is not as convincing as the Medicean satellites. The Tychonian system, kinematically equivalent to Copernicus' but with a geocentered world, also implicitly predicted the correct phases and diameter variations. This isn't at all surprising since the construction is purely geometrical. But there was a high price with this. Still lacking any dynamics, it further divided the behaviors of the celestial bodies by forcing two of them to orbit the Earth while the rest, for unknown reasons, revolved around the Sun alone.

8. The translations in this section are from Grant (1981).

4

FROM THE HEAVENS TO THE EARTH

> You sometimes speak of Gravity as essential and inherent to Matter. Pray do not ascribe that Notion to me, for the cause of Gravity is what I do not pretend to know, and therefore would take more time to consider it.
>
> —Isaac Newton, *2nd Letter to Richard Bentley, 17 Jan. 1692–3*

We now come to the pivotal figure in the history of the concept of force, Isaac Newton (1642–1727). He has been called *the last of the magicians* by Keynes and, indeed, Newton was a person standing between ages. He was a dedicated—obsessive—alchemist and theologian, a hermeticist, and biblical chronologist. But he was also a profound experimentalist, inventor, mathematician, and observer. Although he held a university chair, it was more titulary than faithfully prosecuted. He served as advisor to his government, Master of the Mint, and president of the Royal Society. He was both a theoretical and practical natural philosopher and the paradigm of both in his time. His role was transformational, no area of physical science or mathematics escaped Newton's interests. And none since Aristotle—not even Copernicus, Kepler, Galileo, or Descartes—so completely explicated the mechanisms comprising a single system of the Universe. We will linger in Newton's company for some time and then examine how his physical heritage passed to the next generations in the eighteenth and nineteenth centuries.

THE BEGINNINGS: THE KEPLER PROBLEM

We know from his early jottings, the *Trinity Notebook* and the *Wastebook*, that Newton began to think very early about the motion of the planets, even before the Plague Year of 1665–1666. By his own testimony in his later years:

> In the same year I began to think of gravity extending to ye orb of the Moon and (having found out how to estimate the force with wch globe revolving within a sphere presses the surface of a sphere) from Kepler's rule of the periodical times of the

> Planets being in sesquialternate proportion to their distances from the centres of their Orbs, I deduced that the forces which keep the Planets in their Orbs must reciprocally as the squares of their distances from the centres about wch they revolve: and thereby compared the force requisite to keep the Moon in her Orb with the force of gravity at the surface of the Earth, and found them answer pretty nearly. All this was in the two plague years of 1665–1666.

Distinct from Kepler, however, Newton realized that the force would act in the direction of fall, not in the tangential direction. We will later see this in the second law of motion but for now, we concentrate on how he exploited this insight. If the Moon would move, inertially, in a straight line in the absence of the Earth, then it must be attracted to the center just as any body on the surface is. Ignoring the effect of the Sun to which the Moon is not bound, we can ask what would be the time for fall were the Moon to cease its orbital motion (the origin was which is another problem). The time for fall of any body on the Earth's surface depends on its height, so knowing the distance to the Moon, about 60 Earth radii, Newton calculated the time compared to that needed for a body to reach the center of the planet. This agreed well enough with his hypothesized force law to be encouraging, but not precisely. There are several possible explanations for why the disagreement persisted but the most direct are the incorrect distance to the Moon and the wrong radius for the Earth.

For most of the intervening years between 1666 and 1684, Newton occupied himself with other matters than mechanics. He was deeply engaged in optics, alchemy, and geometrical analysis. His announcement of the prismatic separation, and reconstruction of white light, the discovery of interference effects (known as Newton's rings), the invention of the reflecting telescope, and the systematic examination of optical systems seem far from the preoccupation with celestial dynamics that so deeply engaged him during his student years. But the stage was being set by long, solitary years of extending his experience and meditation.

A chance encounter triggered the revolution. By his own account, Edmund Halley (1656–1742)—with whom Newton was a familiar from the Royal Society—had been studying comets and visited Newton to ask advice on how to treat their motions. On asking what form their paths should take, he received the answer that they should be ellipses. Although, given Kepler's laws of planetary motion this doesn't seem an especially surprising response, Halley reported that he was astonished and, on probing how Newton knew this received the answer "I've calculated it." Halley requested a demonstration and Newton obliged, first with a small account on motion and then, on the urging of both Halley and the Royal Society, embarked on a more comprehensive exposition. It would take him years but, as the historian R. S. Westfall put it, "the idea seized Newton and he was powerless in its grip." The result in 1687, after nearly three years of unceasing effort, was the *Principia*.

THE *PRINCIPIA MATHEMATICAE NATURALIS*

The structure of the *Principia* hides these developmental twists very well and presents the final layout in logical order.[1] Much of the layout of the logic follows from, and supersedes, Descartes' physical treatises. The logical organization of the first chapters of the first book is Euclidean and deductive. Newton opens with the definition of mass, likely the thorniest conceptual issue in the whole work, defining it in effect as the quantity of stuff of which a body is composed. Extension isn't important, as it had been for Descartes. Instead, this is an intrinsic property of material. He also explicitly identified a mass's quantity of motion, what is now called the momentum, with its inertia.

Newton then presents the three fundamental laws of motion. The first states that "Every body perseveres in its state of being at rest or of moving 'uniformly straight forward' except insofar as it is impelled to change its state by forces impressed." This is Newton's statement of inertia, the motion in the absence of any external perturbation. Having defined mass as the fundamental quantity, Newton took the step at the start of creating two separate quantities: the momentum or quantity of motion, and the mass. The second law is that "a change in motion is proportional to the motive force impressed and takes place along the straight line in which the force is impressed." This is Newton's definitive statement of what force means. Two quantities, kinematic and dynamic, are connected. The acceleration produced by any agent is always in the direction of the imposed force. For the first time, this law makes clear that the force can be deduced by its effect, that the *inverse* problem can also be equally well posed—to find the force if the change of state of motion is known. It also introduces the definition of force as a *rate of change* of the state of motion *in time*. Finally comes the third law, that "For every action there is always an opposite and equal reaction; in other words, the actions of two bodies upon each other are always equal and always opposite in direction." This is why Newton required *some* definition of mass because it enters implicitly in the third law. If there is any external agent acting on the system, the reaction of the body depends on the mass—or inertia—and is a phenomenon independent of the composition of the body or the nature of the force. It is always directed in *reaction* to the force and its acceleration is proportional to the mass. Actually, the third law restates the second in a far more general form, asserting the universality of the reaction.

A series of corollaries follows in which Newton deals with the composition of forces in analogy to that of the motions. Since the motions are decomposable into spatial components, those which change display the direction of the force and therefore, *a body acted on by two forces acting jointly describes the diagonal of a parallelogram in the same time in which it would describe the sides if the forces were acting separately*. Colin Maclaurin (1698–1746) in the first popular exposition of the *Principia* for an English speaking audience, *Account of Sir Isaac Newton's Philosophical Discoveries* (1748), expanded this to include the explicit case of decomposition of velocities, which Newton had assumed was already familiar to his readers.

To extend this to the case a of a self-interacting, one in which all the individual masses act mutually on each other, Newton required a modified version of the first corollary, "the quantity of motion, which is determined by adding the motions made in one direction and subtracting the motions made in the opposite direction, is not changed by the bodies acting on one another." This was completed by the proof that would eventually be important for constructing the mechanics of the solar system, that "the common center of gravity of two or more bodies does not change its state whether of motion or of rest as a result of the bodies acting upon each other; and therefore the common center of gravity of all bodies acting upon one another(excluding external actions and impediments) either is at rest or moves uniformly straight forward."

The canvas is already starting to show the big picture. In Bk. I sec. 2 Prop. I., Newton proposes "to find centripetal force." This proposition, the centerpiece of the entire study of gravitation, derives the inverse square for a general motion based only on the *conjunctus* and a continuous curve. But it masks a subtle argument about the continuity of motion itself, not just the geometry of the trajectory, and already introduces the differential calculus that was necessary to connect accelerated motion with its geometric result. In Prop. 5, he seeks to show how, "given, in any place, the velocity with which a body describes a given curve when acted on by forces tending toward a common center, to find that center." Finally comes the moment of triumph, the answer to the question that Halley posed two years before. In Prop. 10, Problem 5, Newton seeks to show that a closed conic section, an ellipse or a circle, is the unique trajectory for a closed motion about a central gravitational source. "Let a body revolve in an ellipse; it is required to find the law of the centripetal force tending toward the center of the ellipse" and in Prop. 11: "Let a body revolve in an ellipse; it is required to find the law of the centripetal force tending toward a focus of the ellipse." With Prop. 17 we find the final statement of the inverse square law and in Prop. 20, Newton solves the general problem of trajectories. This is extended in the next proposition to the hyperbola, and then to the parabola. His answer is the consistency proof that only with an inverse square law can Kepler's orbits be recovered. Then, in the scholium, Newton added that the conic sections are general solutions including the parabola and explicitly citing Galileo's theorem.[2]

Later propositions deal with the determination of orbits from finite numbers of points, a problem later dealt with by Gauss for the determination of orbits from observations. To display the logical structure of the argument, it suffices to outline the rest of the book. The exposition is organized as follows. In Prop. 30–31:, Newton shows the calculation of the orbits, the Kepler equation; in Prop. 32–39: he treats linear descent (that is, freefall); in Prop. 40–42, he generalizes the curvilinear motion to treat arbitrary central force laws; in Prop. 43–45, he shows how perturbations produce a motion of line of apsides[3] of an otherwise closed orbit; Prop 46–56 return to the Galileo problem of pendular and oscillatory motion; Prop. 57–69 expand the application from central forces between bodies to systems of particles, orbits determined by mutual gravitational attraction; and finally Newton presents the solution for the gravitational field of spherical (Prop. 70–84) and non-spherical (but still nested ellipsoidal) bodies (Prop. 85–93).

Here we end our summary of the first book of the *Principia*. Although there are several additional sections, you see both the structure of the argument and the use of the force concept. Newton had, by this point, assembled the machinery necessary for the general construction of the cosmic model of Book III, the *System of the World*. But there were still several points to deal with, especially the refutation of Cartesian physics.

The Banishment of Centrifugal Force

With the realization that motion of the planets can be explained with the unique action of a central force, the need for a *centrifugal* force disappeared. Yet this concept persists in daily life, and it is instructive to see why. Newton adopted a single frame of reference, the Universe at large, and his ideas of space and time were imbedded in his methodology. There is an absolute frame relative to which anything is said to change. This includes velocity as well as acceleration (which, in turn, includes changes in either speed, direction, or both). Thus, for a bucket whirled around a center and attached with a string, any fluid within the bucket appears to be driven to the bottom because it is forced to move relative to this universal coordinate system. For Newton, this was not mysterious, the change in direction was produced by a centripetal force acting through the tension of the string and the trajectory of the contained fluid is constrained by the container. If there were to be a hole in the bottom, the fluid would exit by moving tangentially to the path of the bucket (you might think here of the spin-dry cycle of a washing machine). The material has been torqued so it has angular motion, but the torque need not be maintained to keep the bucket in "orbit," nor do the planets need to be continually pushed *along* their orbits—they only need to be constrained by a radially directed force to remain within the trajectory around the central body.

Now, in contrast, the motion *within* the bucket is *described* very differently. In this frame, we are no longer moving inertially, so there *appear* to be forces. The water doesn't "seek" the walls of the container, instead the walls get in

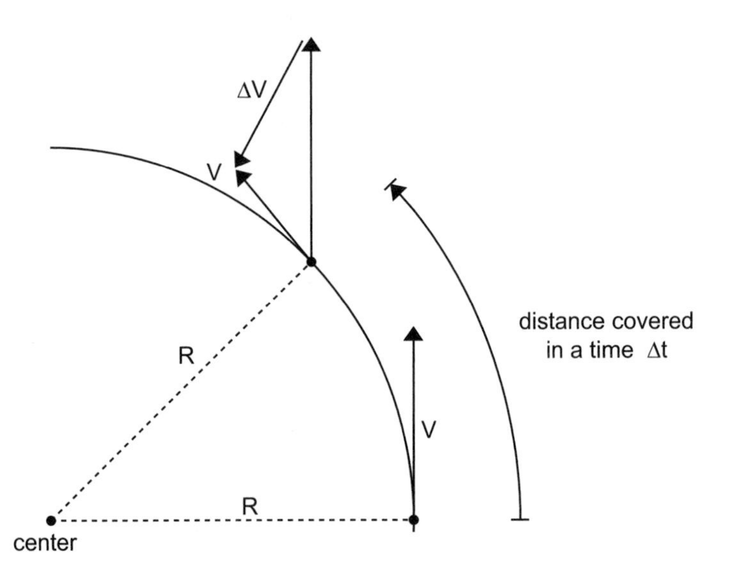

Figure 4.1: The Newtonian construction of centripetal force. A particle moving along a circular path centered at any moment at some distance R has a tangential velocity V. The direction of this motion changes although the speed remains constant. The acceleration, the direction of the change in the tangential component, is toward the center. The force is, therefore, in the same direction according to the second law and equal in magnitude and opposite to the change in the motion by the third.

the way of the water, which is simply continuing to move. The bucket not only confines the motion, it also creates an internal reaction in the fluid that is equal and opposite—in the frame of motion—to that which it would have were a force of equal magnitude to that produced by the motion of the frame applied from outside. The fluid is "completely ignorant" of its frame of motion, as is a person confined within the same moving bucket, so because there is an acceleration there's a "force."[4] It's the laws of motion that permitted the derivation of the universal law Newton sought. The first step is to ask, if we have circular motion, that is the observed acceleration? Moving around a closed curve of constant radius, the direction of motion changes continually. This is just what Huygens had found, but with a basic change of viewpoint: instead of this being the force, it is the acceleration experienced by the test mass, m. For a constant velocity, the body—Newton realized—must be continuously accelerated because it is continually changing direction although not speed. Thus, for a convex curve, the direction of the change always points *toward* the center of the curve. If this is produced by a force, that force must be continuously drawing the body *inward*, hence it is always attractive. With the direction of the change fixed, it follows that the deviation gives a force that points toward the central mass and depends on the quantity of motion mv of the moving body as:

$$(\text{Force}) = -m\frac{v^2}{r}$$

The second law of motion, stated a bit differently, asserts that if we know the acceleration, the force producing it lies in the same direction and has the same magnitude per unit mass. But where this comes from is particularly illuminating. Let's sketch the proof. Imagine a curve with a local center C and take the arc AB to be subtended by an angle ACB and the chord $\bar{AB}$. Take the radius to bisect the chord at some point P. Then at the point of intersection with the arc, Q, connect the chord to form QA and QB. Since the construction is symmetric about the line CQ we'll use only the right hand side. Now since the radius is perpendicular to the bisected chord, $QP : QA = QA : QC$ so we have $QP = QA^2 : QC$. But since QC is the radius to the center R and $QA = V\Delta t$ with V being the speed, then it follows that the deviation of the direction of the motion is the measure of the centripetal force and the quoted law follows. Although this is formally the same as Huygens' result, the direction is opposite: this is a force *toward* the center C, not a *conatus* produced by a centrifugal tendency. While the result of this law had already been known 20 years before Newton's work, it was the *Principia* that, by reversing the direction, forever changed the concept of gravitation as a force. Yes, this seems rather flowery prose but I know no other way to imagine what it might have felt like to have this insight and you would do well to contemplate its significance.

The force must increase with increasing mass of the central body so the acceleration it produces is independent of either the mover's mass or time. Furthermore, the force must act only toward the center of the body. In fact, even if each parcel of mass attracts the body, only that toward the center is needed. We could imagine a body so distant from the mover that its size is essentially that of a point. Then the proposition is simple, we need only consider the radial distance. For a spherical body of finite size, the portion closer to the mover has a greater attraction but, for a fixed solid angle, there is less of it, while that at greater distance exerts a weaker attraction but there is more of it. In the end, *these sum to the force all of the mass would produce if all of the mass were concentrated at the precise center of the body.* To extend this, Newton introduced what we could call the interior solution, the one that would be exploited later by Laplace and Poisson: for a homogeneous set of nested spheroids—without lumps so they are symmetric—*only the mass interior to any radius attracts the test mass and any external matter produces a net cancellation.* Again, this is an effect of summing (integrating) over surfaces of fixed solid angle. We can step back, somewhat anachronistically, and write that dimensionally the acceleration is R/t^2 (where here R and t are intervals of space and time, respectively) so the force must vary as $1/R^2$ since the velocity is R/t. Put more directly, since we know from Kepler's harmonic law that the orbital period varies such that R^3/t^2 is constant and the same for all orbiting bodies we arrive at the same result. This had been the first route Newton had taken, comparing the acceleration of the moon with that of a falling mass on Earth scaled to the lunar distance and found good agreement, at least good enough to pursue the argument further. By this point in the writing of the *Principia*, twenty years later, he knew the answer.

Universality

At the heart of the *Principia* lies a radically new conception of the Universe. Not only is it governed by physical laws, or natural compulsions, it is all the same. The structure and operation of the world can be understood by a *simple, universal law*, gravitation, that not only explains the phenomena but predicts new ones. How different this is from the medieval world. This is not from a hypothesis. It is from a theoretical generalization of empirical results that now are predicted to be consequences of the laws. The motion of the planets is the same as terrestrial falling bodies. The tides have yielded to a single force that is the same one responsible for the motion of the Moon. There is no need to posit a mechanism to transfer the force, although it seemed Newton couldn't restrain himself any more than others in seeking one out, nor do we need to understand the source of this force. The methodology is now completely different, we will hence derive the properties of matter from the observation of interactions by applying a few laws and then deriving their effects.

To understand the revolutionary calculation contained in Book I of *Principia* consider how the the laws of motion have been applied. The solution to circular motion, more than any other proposition in *Principia*, shows how dramatic is the effect of Newton's laws of motion. The change in the fundamental picture comes from the role of inertia. Were a body to accelerate *centrifugally*, that is, attempting to recede from the center of motion, its behavior would have to be an intrinsic property of matter. Instead, by fully grasping what inertia means, Newton truly dispensed with any such *occult properties* of matter, in the very sense of the criticism later leveled against him by Leibniz. The body does not seek to move anywhere other than to preserve—notice, this is not *conserve*—its state of motion.

With this as the definition of inertia from the first law, with no explanation either required or given since he affirms it as a law, Newton was able to explain the *apparent* centrifugal action by a simple change in reference frame. This is not how it is described in the propositions of Book I but that presentation amounts to the same statement. Instead, taking the acceleration to be the deviation *toward* the central body relative to *tangential* motion, which is the instantaneous continuation of the inertial motion, then the force becomes radial and acts only between the centers of the masses.

The striking result is that the answer is identical. Only the sign is changed for the force. But what is far more important is that this is a general result, not one confined only to circular motion: since the mover is deviated at every instant, and the limit can be taken relative to a continuous curve as Newton showed in Prop. 8, there is no restriction. Any conic section can result depending only on the initial conditions. Since the forces are only between the bodies and neither external to the pair nor tangential, as he had shown extending the treatment from Galileo and from ballistics, the limiting case for a constant acceleration will be a parabola and therefore any eccentricity can result. Recall that the limiting case for a critical horizontal velocity, that the radius of the orbit is the same as the radius of the Earth, is an interesting variant on the Galilean argument. Since for a circular orbit the distance from the gravitational source is constant, this is actually motion under *constant acceleration* and shows that the limit of the parabola or ellipse going to the circle depends on the tangential speed alone.

Newton's illustration of the limiting orbit for a cannonball fired horizontally to reach a limiting speed is a spectacular graphic as well as a precise statement of the problem, one that can easily be extended to an escape condition is the initial tangential velocity exceeds the orbital limit. This is, however, not enough to explain the motion of the Moon or any other pair of orbiting celestial bodies, including the Earth–Sun pair. While the first and second law suffice, without modification, to yield the solution for an enormous mass ratio between the mover and central body, the third law also provides its extension to finite, near unity, mass ratio. By saying that the force is mutual and opposite, that the two bodies experience equal and opposite forces but consequently accelerations in inverse proportion to

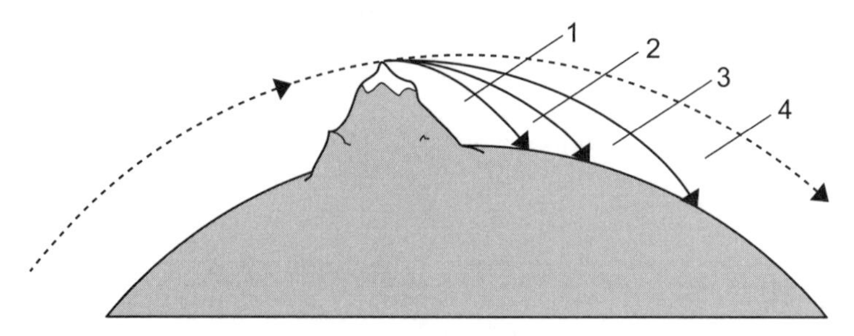

Figure 4.2: Newton's example for orbital motion from the *Principia*. A projectile is launched with increasing initial horizontal speeds. Eventually there is a critical value at which the body falls at the same rate as it displaces horizontally, maintaining a constant distance. Thus circular motion is always accelerated—it's always falling but never hitting the ground.

their masses, Newton made the critical step required for a celestial mechanics. The center of mass, or center of gravity, remains unmoved in the absence of an external force. For the Earth–Moon system, this means the two bodies, if stopped in their orbits, would fall toward a fixed point in space. Notably this seems to endow space itself with a special property, one that was later to exercise Newton's ingenuity and later commentators in much the way the discussion of space and place was central to Aristotle's thinking. The existence of this point is another expression of inertia, that the moving system has itself a common center that will have its own trajectory if a third body or external force acts. So for the Earth–Moon–Sun system, this becomes the explanation for the continued binding of the Earth and Moon while their common center orbits the Sun. The same is true for the other planets and their respective moons and rings.

The presence of this external force, in the Earth–Moon case, also disturbs the lunar orbit. While various deviations in the orbit had been known from the early modern period, and even Ptolemy was constrained to add special kinematic device—the moving equant—to account for the precession of the apsidal line of the lunar orbit, Newton had discovered its dynamical explanation. The relative magnitudes of the forces of the Earth and Sun on the lunar motion reduced the solar influence to a mere perturbation. Were it stronger, the Moon could instead orbit the Sun and be perturbed by the Earth as would happen with any of the other satellite systems. But with this realization, the scaling of the solar mass and, therefore, the density of the two principal bodies affecting the Moon could be determined and this is included in Book 3 of the *Principia*. We will return to this shortly.

The logic of Newton's argument is then clearer if we think to give this notion of a center of mass a prime role in his thinking. First find what we would now call a reference orbit and then, after demonstrating that it must be one of the two closed conic sections, show how it will be disturbed in the presence of any other gravitating body, i.e., mass. By the end of the first book the reader is really prepared for Book 3, the System of the World, which is the applied part of the *Principia*. But before taking such a step, and in particular because the Cartesian system of fluid driving in the place of his gravitational force was so widely diffused in the philosophical community of his time, Newton felt compelled to not only examine the application of the force law to fluids but also to use it to refute the vortex picture. The study is as important to his system as it is to demonstrate the enormous reach and power of the new mechanics.

In Book I, Newton solved the two principal problems of celestial mechanics, at least to a first approximation. One deals with the motion of the center of mass of a two body system in the presence of perturbers, the main issue in the stability of the Earth–Moon system in orbit around the Sun but also the central issue regarding the overall stability of the Solar system. He showed that the centers of mass of the two systems can be treated as pseudo-mass points whose motions are determined by the relative intensities of the forces of the nearby—and distant—masses of whatever magnitude. The second, far more subtle problem is related to extended masses: what is the gravitational force of an extended body, that is when the distance between the masses is comparable to their size? He demonstrated, in a series of propositions accompanied by examples, that the attraction on a point mass is the same as if the total mass of the more extended sphere—the one case he could resolve directly with a minimum of approximation—is concentrated at its center. To go to progressively more realistic problems, he then showed that the same is true for two spheres, whatever their relative density distributions, thereby permitting the solution of the Kepler problem even when the two objects are very large. Finally, and this is perhaps the most beautiful of the results at the end of the first book, Newton showed that the attraction of the mass *within* a sphere on its superficial layers is as if the mass is concentrated at the center and *only* the mass within a given radius attracts the overlying layers.

This is one of the few examples in *Principia* where Newton resorted to symmetry arguments. He began with a hollow shell, an idealization that allowed him to use the inverse square law explicitly to show that the force cancels exactly within the sphere for a mass placed anywhere inside. His argument is simple and elegant. The solid angle subtended by any piece of the shell is constant but the mass within that cone depends on the area. Since that varies as the square of the distance, while the force produced by the shell varies as the inverse square of the same distance, the two intersecting cones produce equal and exactly oppositely directed forces and therefore cancel. The same argument served, with the asymmetry of the test mass being situated outside the sphere, in the later propositions that demonstrate that the force is as if all the mass were concentrated precisely at the center of the body. Then, taking two differently stratified spheres, he extended this to two finite bodies. But the *tour du force* came with Newton's demonstration that this same argument also holds for nested *homeoids*, concentric spheroids with precisely the same symmetry. Thus for any density distribution it was possible now to derive the gravitational force on any internal shell. With this, he passed from the dynamical to the static problem and introduced the line of research that opened the structure of the Earth, and of all celestial bodies, to remote sensing. With this construction, it was possible to study the inverse problem although, as we will see, that took one more conceptual change to complete the picture.

A weakness of Newton's calculation was realized and corrected rather soon and this was an essential extension to the theory. Imagine a pointlike mass in the vicinity of an extended body. Then by linear superposition, the fact that we can simply add the gravitational fields produced by individual masses each in its appropriate direction, each piece of the body attracts the mass (and vice versa) according to its distance. Thus although the center of mass is precisely defined with respect to the two bodies, the resulting gravitational field depends differently on the between the bodies than a simple inverse square law. This is because some parts of the body are closer than others and also some are farther from the line of centers. The resulting representation is a series of polynomials in the two relevant angles, the azimuthal and the meridional, named after their originator Legendre. It was, however, Laplace who saw the generality of their properties and showed that these are the solutions to the gravitational field for an arbitrary body when expressed in spherical coordinates. An orbit, even in the two body problem, is never a closed simple ellipse. For any body of finite radius, R, there are always terms of order $(R/r)^n$ for any arbitrary n and distance r. What Legendre had shown was that "his" epiphanous polynomials resemble a distribution of point masses acting *as if* they are dipoles, quadrupoles, and higher order. Thus, for instance, a bar is a sort of dipole although with two equal "charges."

Refutation of Cartesian Methodology

After its concentration on orbital mechanics, it might seem strange that the next topic addressed in the *Principia* is the motion of a continuous, fluid medium and its action on internally moving bodies. Yet you should recall here the audience to whom the work was addressed. The subject occupies virtually the whole of the second book. It was one thing to have treated the behavior of things, but *causes* were lacking for the forces, especially gravity. While various hypothetical explanations could be supplied for most material, contact forces, gravity resisted. For one thing, acting apparently at a distance, gravitation required some means for transmission, yet nowhere in the first edition does Newton explicitly say anything about this. As we will discuss later, before the General Scholium, added to the second edition and refined for the third, there was nothing to satisfy this need for theorizing about sources. But there *was* already an "explanation" for this universal force—the Cartesian plenum, and for its centripetal action—the vortex model, and it was this Newton sought to quash with his analysis of motion of bodies in continuous media. The treatment is general, without specific reference to the cosmic medium.

Coming as it does after the full presentation of the mechanics of particles and massive finite bodies, Book 2 of *Principia* was the explicit response to and refutation of Descartes' qualitative picture of the fluid-filled universe. If the fluid were viscous, the moving masses would experience a drag that would change their orbits over time. This was a negation, from an unusual direction, of the Cartesian viewpoint. A vortex could indeed explain the circulatory motions, although it would apply only for circular orbits. An elliptical vortex is inexplicable, and we now know not even stable. Newton also demonstrated that the velocity law for a central vortex could not produce the known law of planetary motion. To make matters even more difficult to arrange, nesting these was a nearly impossible task. For a universe filled with such motions, each center was the source of gravity.

There were several reasons for this frontal assault on the vortex picture. The first was philosophical. The *Principia* is really one long argument[5] showing how the application of the concepts of inertia and force, along with the affirmation of the universality of gravitation acting on all masses in the universe, both explains the existing phenomena and predicts new ones. But there was a sub-text that remained even more important: whatever gravity *is* isn't important. This point was driven home in the General Scholium that Newton added to Book 3 when revising the book for the second edition. There he explains what had been only implied in Book 2, that it isn't necessary to understand the ultimate causes (even though Newton had his own ideas about them), it was enough to understand the mechanisms. The Cartesian picture was a qualitative, not quantitative, picture of how things work but it was also a proposal for what things *are*. The end of Book 2 contains an explicit, biting scholium summing up the whole case against the vortex theory of planetary motion, ending with the statement:

> Therefore the hypothesis of vortices can in no way be reconciled with astronomical phenomena and serves less to clarify the celestial motions than to obscure them. But how these motions are performed in free space without vortices can be understood from Book 1 and will now be shown more fully in Book 3 on the system of the world.

Figure 4.3: Figure from Descartes' *Principia* showing his explanation for gravitation based on a universal fluid filled with contiguous vortices. Each star is the center of a vortex. This is the world later described by Voltaire and contrasted with the Newtonian vacuum. It was this model for planetary motion that spurred Newton's composition of the second book of his *Principia*. Image copyright History of Science Collections, University of Oklahoma Libraries.

An indication of the success of the second book is the remark by Huygens, originally among the most dedicated Cartesians, that vortices disappeared in the strong light of Newton's demonstration of their failure to produce the Keplerian laws of planetary motion. Most later commentators understood the point as well. But there were those for whom the Newtonian concept of gravitation didn't suffice to provide a complete, coherent philosophical construction. One of them was Bernard Fontenelle and it's worth a moment to reflect on his reaction to the *Principia* to see how strange the new concept of force was to Newton's contemporaries.

In his popularization of the Cartesian system in 1686 (and many later editions), Fontenelle extended the vortices to fill the universe, imagining each star the center of its own system with a multiplicity of inhabited worlds. By coincidence, the work, *Conversations on the Plurality of Worlds*, appeared nearly simultaneously with the first edition of the *Principia*. The work was immensely popular on the continent, more accessible than Descartes or, later, Newton, and written with a high literary style and and lightness of presentation. It was a serious work, and consequently also a serious challenge for the view that space is an empty immensity in which gravity acts by principle without a mechanical explanation. For the literate public, Fontenelle was both a port of entry into the new philosophy and a guide through the newly discovered territory. His popularizations of the activities of the *Academie des Sciences* was the first *yearbook of science*, and was widely distributed and read on the Continent. Dedicated to this mechanical explanation for gravitation, his last word on the topic was published anonymously in 1752 and again championed the Cartesian system nearly 30 years after Newton's death. There is nothing in this picture that requires the vortices; it was only Fontenelle's choice to have a physical explanation behind the mechanism.

But there were other reasons for Newton's digression on continuous media and fluid mechanics in Book 2. Remember that the pendulum was used for measuring not only time but also the local gravitational acceleration.[6] Since Galileo's

Figure 4.4: Isaac Newton. Image copyright History of Science Collections, University of Oklahoma Libraries.

discovery of the isochronism of the pendulum, a simple thread and bob provided a standard a fundamental measure of time at any place. But the comparison between sites requires a very precise calibration and this can only be done knowing how to treat both the friction of air and the increase in the effective mass of the bob when moving through the background. This would later be indispensable in the measurement of the arc of latitude of the Earth and the study of its gravitational field. In addition, the motion of a body in air, instead of a vacuum, was a mechanical problem of some interest, especially for ballistics, and the complete solution for the dynamical resistance of air was quite useful and an important extension of the problems treated in the first book. It was another way of seeing what inertia means because, since the frictional and resistive forces of a surrounding medium depend on the properties of that medium (for instance, its density and temperature), its impedance of the motion of a projectile dissipates the impulse. This had been seen completely differently in medieval mechanics: in the impetus theory, you'll recall that the medium actively drives the motion that would otherwise decrease with increasing distance from the source. In addition, Newton hypothesized that the force of the medium depends on the velocity of the body, hence unlike gravitation or contact forces it decreases as the body slows down and is independent of the distance it covers.

The System of the World

The first two books of the *Principia* were methodological and demonstrative. The third contained the applications, but not, as they had been for Galileo, to problems of terrestrial mechanics. Newton had a much vaster territory in mind, the whole universe. And to show the power of his new conception of gravitation nothing could serve better than the explanation of some of the most perplexing natural phenomena that had remained problems since antiquity. The tides, precession of the equinoxes, and the stability of orbital motion against perturbations were the subjects of the third book of the *Principia*, the promised *The System of the World*. Here Newton dealt with the particulars, the demonstrations of the explanatory and technical power of this new physics: its application to the fundamental questions of astronomy.

Try to imagine what it must have felt like, to see the whole of Creation open to analysis, to sense the evolution of everything, and to have a conceptual and computational battery with which to accomplish the task. A century later this was the inspiration for poets, as William Wordsworth wrote in his *The Prelude* (1805),

> And from my pillow, looking forth by light
> Of moon or favouring stars, I could behold
> The antechapel where the statue stood
> Of Newton with his prism and silent face,
> The marble index of a mind for ever
> Voyaging through strange seas of Thought, alone.

The tides provided one of the earliest successes for the Newtonians. Its clearest presentation is, however, not Newton's own but the 1696 memoire by Halley for the Royal Society. Newton showed (Cor. 2) that the center of gravity—(*i.e. center of mass*, here gravity has the same meaning as it would have to Galileo)—moves with a fixed angular momentum and in a closed orbit, experiencing only the centripetal force of the attracting mass. But if we have an extended ensemble of bodies, which are mutually attracting and also attracted to the central mass, those at larger distance cannot have the same orbital period as those closer to the center. Since they are forced to move with the same velocity as the center of mass at distance r from the central body, in the orbiting frame their centrifugal acceleration compared to the gravitational attraction of M produces a differential acceleration. The body will distend along the line of centers to balance this excess while keeping the same motion of the center of mass. The oceans remain in overall hydrostatic balance (this was the second time hydraulics was invoked on so large a scale, the first was Newton's argument about the equilibrium of channels and the shape of the Earth in Book II) but rise by a small amount in instantaneous response to the change in this differential acceleration. The solid Earth transports the fluid and the *static* response, independent of any induced flows, causes a twice daily, or semi-diurnal, change. The bulge of the fluid is coupled to the rotation of the Earth and this mass produces an *equal and opposite reaction* (again, the role of the third law) on the perturber. In this case, the Sun is so much more massive that the effect on its structure and motion is insignificant.

For the Moon, the situation is complicated by the difference between the periods of lunar orbital motion and terrestrial rotation. The Moon moves about 1/30 of its orbit per revolution of the Earth, in the same direction, so it always leads the Earth *were the Earth to stay motionless*. But since the rotation period is one day—just saying this the other way around—the bulge is always carried forward and this extra mass is now always *ahead* of the instantaneous position of the Moon. Being so low mass, the Moon receives a greater relative acceleration than the Sun and moves outward in its orbit. The Earth, as a reaction, spins down. This produces two effects, a steady lengthening of the *solar day* and a lengthening of the period of the Moon's orbit.

THE GENERAL SCHOLIUM

On the urging of Roger Cotes, his editor for the second edition of the *Principia*, Newton was induced to add something about his theology to counter charges of implicit atheism in the work. It is hard to see now why such an orthodoxy was required. The statements in many of Newton's writings, such as his correspondence with Richard Bentley regarding the implications of the new physics for religion, certainly attest to a deep conviction in the direct action of God in the world. But somehow the system was incomplete and through this qualitative and extremely carefully worded gloss on the whole of the work, Newton also touched on one of the fundamental cosmological questions: the stability of the system.[7]

> Hitherto we have explain'd the phaenomena of the heavens and of our sea, by the power of Gravity, but have not yet assign'd the cause of this power. This is certain, that it must proceed from a cause that penetrates to the very centers of the Sun and Planets, without suffering the least diminution of its force; that operates, not according to the quantity of surfaces of the particles upon which it acts, (as mechanical causes use to do,) but according to the quantity of the solid matter which they contain, and propagates its virtue on all sides, to immense distances, decreasing always in the duplicate proportion of the distances. Gravitation towards the Sun, is made up out of the gravitations towards the several particles of which the body of the Sun is compos'd; and in receding from the Sun, decreases accurately in the duplicate proportion of the distances, as far as the orb of Saturn, as evidently appears from the quiescence of the aphelions of the Planets; nay, and even to the remotest aphelions of the Comets, if those aphelions are also quiescent. [***But hitherto I have not been able to discover the cause of those properties of gravity from phaenomena, and I frame no hypotheses. For whatever is not deduc'd from the phaenomena, is to be called an hypothesis; and hypotheses, whether metaphysical or physical, whether of occult qualities or mechanical, have no place in experimental philosophy. In this philosophy particular propositions are inferr'd from the phaenomena, and afterwards render'd general by induction. Thus it was that the impenetrability, the mobility, and the impulsive force of bodies, and the laws of motion and of gravitation, were discovered. And to us it is enough, that gravity does really exist, and act according to the laws which we have explained, and abundantly serves to account for all the motions of the celestial bodies, and of our sea.***]

> And now we might add something concerning a certain most subtle Spirit, which pervades and lies hid in all gross bodies; by the force and action of which Spirit, the particles of bodies mutually attract one another at near distances, and cohere, if contiguous; and electric bodies operate to greater distances, as well repelling as attracting the neighbouring corpuscles; and light is emitted, reflected, refracted, inflected, and heats bodies; and all sensation is excited, and the members of animal bodies move at the command of the will, namely, by the vibrations of this Spirit, mutually propagated along the solid filaments of the nerves, from the outward organs of sense to the brain, and from the brain into the muscles. But these are things that cannot be explain'd in few words, nor are we furnish'd with that sufficiency of experiments which is required to an accurate determination and demonstration of the laws by which this electric and elastic spirit operates.

Thus, in the General Scholium, although he took great pains to make his philosophical (more to the point, theological) position clear, Newton flatly refused—in any edition of the *Principia*—to enter into a debate on the *cause* of gravity, asserting that it is merely an intrinsic property of matter. This was not his problem, nor did he recognize any need to seek its cause. Although he seemed to almost echo an unlikely predecessor, Ptolemy, who considered that the aim of a model is to reproduce the appearances of the heavens—to *account* for their motions—Newton's physics went much farther. He required complete consistency among all parts of the explanations of the motions and only avoided questions of *ultimate* causes. Yet he couldn't completely avoid the issue, not only because of his audience's expectations but also because of his own theological preoccupations. The Prime Mover had been eliminated as the source for the motion but God was now required to maintain the system in a state of perpetual motion and stability. So in the General Scholium Newton included the seemingly enigmatic declaration "hypothesis non fingo," *I frame no hypotheses* but the context of this affirmation was unambiguous. He lacked any causal explanation for gravitation, postulating it as a universal force that acted between bodies only by virtue of their mass. In this, his construction of a physical picture is completely concordant with the modern program of physical science. By postulating an attribute and providing a formal mathematical representation of its force law, the direct problem of motion can be solved within a consistent set of axioms, derived from experience. The details of the interactions, how the force is transferred between bodies for instance, doesn't matter as long as the mathematical law for its action can be prescribed. There is no need to postulate *final* causes in the original sense of that word, the ultimate origin of these physical laws, to make progress.

The universe becomes spatially unbounded in Newton's final conception. In order to preserve its stability, he demonstrates the enormous distances of the stars and that their cumulative forces should be very small. Homogeneity of the overall distribution is required, any excess would become self-gravitating and attractive to all masses in its immediate vicinity. Given enough time, this should lead to a complete collapse of the entire system.[8] His provisional solution was theological, that the system is occasionally "reset" from the outside by God to maintain the state of the universe.

In summary, even without a cause of gravity, the inheritance of the Newtonian revolution was an intoxicating the vision of a truly self-consistent natural philosophy that could explain all phenomena. The successes in celestial mechanics, in fluid dynamics, and in the gas laws were truly impressive. But while mechanics was reformed by the Newtonian methodology, it took almost two hundred years to explore its richness and firm up its foundations.

THE NEW CELESTIAL MECHANICS

Planetary motion was the inspiration for, and critical test of, Newtonian gravitation theory. Solving the problem of the orbit for two isolated bodies, the basic Kepler problem, was sufficient to demonstrate the success of the new concept of a universal force acting between masses at a distance but not enough to extend that theory to the construction of the world. The success of the Newtonian concept of force and the inertia of circular motion removed the need for tangential drivers, which you'll remember even Kepler required, and provided the means to geometrically model and calculate the actions of forces in arbitrary arrangements of masses. we've seen, since an acceleration *requires* a force as its cause, and the accelerations are the observables in celestial mechanics in ways more obvious than anything else in Nature, the forces can, *in principle*, be deduced by a physical procedure. But what does that mean? When you observe a planet's position, it is with respect to the stars whose positions provide a fixed reference frame. The lunar theory was the first success of the new celestial mechanics so let's examine how that evolved as a way to understand this statement. The month is the mean motion of the Moon against the stars, the circular orbit that should result from motion around a gravitating center of mass. But this constant period doesn't suffice. The rate of motion varies with the position in the orbit and the season of the year. A purely kinematic approach takes these and assigns each a "correction," without explaining the origin, so you have at the end a complex set of rules. These can be simply encoded as a set of tables that depend on a number of variables: the time of the month, the time of the year, or the phase in the motion relative to some fixed point. These are the deviations. With the Keplerian ellipse the most important are immediately accounted for, since the motion is *not* actually circular the effect is to alter the angular displacement depending on the phase *in the orbit* relative to the reference position independent of the time. There is a choice here, one can use the orbital phase or the time after perigee passage, they're equivalent. The same construction, with less success, had been attempted by the introduction of the eccentric in the geocentric picture for the Sun and planets, for the Moon it was in a sense simpler because it *is really* going around the Earth. The difference comes when attempting to predict phenomena such as phases of the illumination or eclipses. Then you need to solve the motion *relative to a moving object*, the Sun, and this requires a second orbit. It was this which produced the difficulties since the phase of the perigee, *relative to the solar motion* changes in time. Thus the time, not only the angle in the orbit, is required. It is still possible to refer everything to a circular orbit, however, or to a Keplerian ellipse, and treat each successive alteration purely geometrically to obtain the successive approximations to the exact motion and predict the times of eclipses and occultations as well as phases.

Even by the middle of the seventeenth century it was not clear that the changes in speed required a force and that agent could depend on the distance. This is perhaps best seen in Galileo's misapprehension of the cause of the tides. The problem of freefall is fundamentally different in appearance than that of planetary motion because it doesn't require a cause for the acceleration, at least not a precise law. The law for gravitation was not Galileo's concern. He was dealing with forces of a different sort, and especially the general notion of inertia and equilibrium of bodies under mutual action. The rest would remain speculation. What kept the Medicean stars moving around Jupiter, or cause the companions of Saturn to move with the planet, was not *his* question, that it happened was enough to demonstrate that a center of circular motion could itself move. Thus the Moon could stay attached, somehow, to the Earth while it orbits the Sun. Within the Newtonian approach, however, any change in the angle relative to the central force orbit, the reference ellipse, requires an additional perturbation. All changes in motion are along the lines of the forces, thus all orbits require centripetal forces. The proposition thus allows one to discover the origin of every irregularity by determining the sense, at any instant, of the change in the state of motion. The linear superposition of the forces in the same direction and parallelogram law for their addition if they are not parallel or even collinear, both demonstrated in Book I of the *Principia*, suffices to insure that once an acceleration is known so is the magnitude and direction of the force that produces it. Thus we see the tremendous power of the laws of motion as Newton stated them in the celestial context. The first states "look at the motion, see if it is changing in direction and/or magnitude; if it is, then there is an external cause." The second continues "and, by knowing the acceleration you know the direction of the cause," finishing with the third law asserting "and the magnitude of the mass multiplied by the acceleration is directly equal, for any quantity of mass and irrespective of its composition, to the force producing the change." The inverse problem can be solved once you can make a kinematic measurement, to reason from effects back to causes without needing additional hypothetical constructions.

These variations are accelerations caused by forces that act along the lines of centers to the other bodies. Thus every variation from a perfectly centrally symmetric motion requires an additional force "agent" or property of the body, just a different center of attraction. For instance, the solar mass is large enough to require including its attraction in the motion of the Moon even though the distance is enormous, and this is one of the causes for the orbital precession and also of the tides. If it causes one because of the finite size of the orbit of the Moon, it produces the other in a similar way by the finite size of the Earth. *One unique explanation suffices for a host of apparently distinct phenomena*. The additional proof that the gravitational attraction is central only if the body is spherical and that there can be additional accelerations produced by a finite, nonspherical body, extended the force law to treat much more complex problems, eventually even the motion of satellites in the near field of a nonspherical planet.

Remember that Newton never, throughout the long discussion in Book I, specified what gravitation *is*. How it is transmitted between bodies was left to speculation, of which there were many in the subsequent years. He faced the criticism of the Cartesians who, even in the absence of quantitative results, propounded a physical origin for the field that Leibniz called an *occult quality*—attraction—that was universal and external to the body. Whatever his conceptualization of its mechanical origin, Newton's program was to demonstrate how it unified and resolved fundamental problems and rendered calculable things that had previously been treated only phenomenologically. Even if he didn't solve many of e problems he posed, in particular those of most importance to the calculation of the lunar orbit such as the variations in latitude and the tides, Newton succeeded in setting out the framework in which such problems *could* be solved with eventual development of the computational machinery. These results had a dramatic effect on Newton's contemporaries and framed the research program in celestial mechanics for nearly two centuries until the possibility of non-gravitational forces was introduced at the end of the nineteenth century. Again, we can see why. The single source for the accelerations reduced the physical complexity of the problem. Only the form of the law, not its ultimate origin, is needed along with the equations of motion. The conservation laws reduce to only that of inertia. There is no need to separately treat angular and radial momentum, a tangential force resulting from the presence of another body than the principal center of attraction or any nonspherical central or external mass distribution will produce a change that is identical to a torque. In the two body problem this is the simple explanation for the stability of the direction of the line of apsides (if there is nothing to break central symmetry an orbit once fixed in space remains in the same orientation) and the constancy of the eccentricity. When in Book II Newton digresses at length on the dynamics and properties of fluids, and in the third book asserts that vortices are completely unnecessary to explain the planetary motions, he understood that in a vacuum there is no drag (hence the linear momentum remains constant subject only to gravitational driving) and that the tangential motions were imposed at the start of the world and thereafter required no additional force to maintain them.

The first *predictive*, as opposed to *explanatory*, success of the new mechanics was the return of the comet of 1682, the comet that initiated the exchange with Halley that finally impelled Newton to compose more than the short tract *de Motu* and produced the *Principia*. It was also the demonstration of the universality of gravitation. Comets presented a particular difficulty for astronomers. They had been included in Aristotle's *Meterology* as terrestrial phenomena, notwithstanding their affinity to the heavens because of their diurnal motion with the stars. But their appearances were random, both in timing and location, and their motions were never as regular as those of the planets. Furthermore, they are obviously very different bodies, nebulous and irregular, moving both within and far from the ecliptic. With Georg von Peuerbach's failure, in 1457, to detect a cometary parallax and Tycho Brahe's solution for the orbit of the comet of 1577 that required a radial motion that crossed the spheres, these bodies were obviously beyond the atmosphere. Descartes explained their paths as complicated weaving between planetary vortices, and Galileo and Kepler had also dealt with their appearances and motions. But with the apparition of an especially bright new comet in 1682, Halley noted the regularity of historical comets with intervals of about 76 years and hit on the possibility that this was the same body as that seen in 1607 and earlier. Not only was it superlunary but it was also recurrent. Armed ultimately with the machinery of the new gravitational theory, he determined its orbit as a conic section, an ellipse, and predicted its the return in 1757. When first detected on 25 December 1758, it proved a sensational triumph for the new mechanics and spawned a mania to search for, and determine the properties of, new comets. Since the orbit is very elliptical, and the aphelion distance is so large that it reaches beyond the then-known planets, the orbit itself extended the reach of tests of gravitation to about 30 times the distance of the Earth from the Sun, past the region covered by the theory presented by Newton.

Similarly, a cosmical test for gravitation came with the determination of the relative masses and densities of the planets. It was possible as early as 1748 for Maclaurin to present a table of the relative constitution and properties of the major bodies in the solar system, Jupiter and Saturn on account of their moons and the Sun by the period of Mercury scaled to that of the Moon, in terms of Earth masses. Thus the solar perturbing contribution to the lunar orbit could be established and with that, the motion of the Moon could be predicted with fair accuracy. With this determination of the masses of Jupiter and Saturn and a knowledge of their reference orbits, the anomalies in their motions were also explicable. One of the first challenges was to solve the motion of these planets under their mutual perturbation. In a prize submission to the Paris Academie in 1743, Leonard Euler showed that the motion is stable under periodic perturbations This became known as the *Great Inequality*, the near resonance of the two orbits in a 5:2 ratio of periods (Saturn to Jupiter). The near resonance means that with the passage of time, the force becomes quite large because the perturbation remains almost stationary in the moving frame and never cancel on average. Any resonance (integer ratio of the periods) causes serious, potentially fatally disruptive problems for orbital motion under any sort of field, or even a harmonic oscillator.

The three body problem isn't merely the addition of another mass, it requires solving the equations of motion in a rotating frame. Formulated to address the lunar problem, in which the principal masses (the Earth and Sun) have an orbital period of one year and small eccentricity, the problem was only made tractable with mathematical advances and not resolved fully until the twentieth century. But in the mid-eighteenth century, it had become a serious challenge to the stability of the world. The fundamental problem was set out in Book I of *Principia* and elaborated in Book 3. Pierre-Simon Laplace (1749–1827) finally succeeded in showing that the perturbations were periodic on a very long timescale and, therefore, although the motion might appear now to be changing in only one sense (that is, secular), over a greater interval of time the orbits are stable. Lagrange demonstrated that there are a set of fixed points in the orbit of a third body of negligible mass (relative to the principal components) where perturbations leave the mass unaltered. In the so-called restricted problem, in which the eccentricity of the main binary is vanishingly small and the orbital period is constant, there are two equilibrium points (where the attractive force is balanced by the inertia of the particle through the *centrifugal* force that appears in the non-inertial rotating frame. These lie at the vertices of equilateral triangles and are along the line perpendicular to the line of centers of the binary on opposite sides of the orbit but in the same plane. There are three more, where the force again locally vanishes, but these are unstable and lie along the line of the principal binary. The discovery of the Trojan asteroids, at 60 degrees relative to the Sun in the orbit of Jupiter demonstrated the existence of the L_4 and L_5 points.

The discovery of Neptune in 1846 was the most celebrate result of celestial mechanics of the nineteenth century.[9] Following its discovery, Uranus had nearly completed an orbit at the end of the 1830s by which time the secular departure from its predicted path was significant. While this could be merely the result of accumulated bad positions, the astrometry was sufficiently precise to afford an alternative, and more plausible, explanation: the orbit was being gravitationally perturbed by another, yet unknown, planet. By that time, the masses of the principal planets are known from their moon systems. Since each perturbation produces not just a displacement but an acceleration (the angular motion changes in time), in the now standard procedure of the inverse problem the location of the perturber could be obtained by noting the time dependence of the changes in Uranus' motion. If the perturber is stationary, the reference orbit won't close and precesses. Instead, if the perturber is moving, the axis between the bodies will follow the motion of the unknown orbiter. The equations of motion, though nonlinear and therefore almost impossible to solve in closed analytic form except in very special approximations, can nonetheless be integrated numerically since they depend only on the initial conditions. Two solutions were found, by John Couch Adams (1819–1892) and Urbain Jean Joseph Le Verrier (1811–1877), both similar, but the latter was the basis of the discovery of the planet by Galle at the Berlin observatory on 23 September 1846. In the most stringent test, on a cosmic scale, celestial mechanics was thus seen to have spectacularly confirmed and firmly established the description of force in general, and gravitation in particular, as it had developed during the previous 150 years. Throughout the nineteenth century an increasingly accurate and complex set of algorithms were developed for just this problem, how to find the future orbits of the bodies of the solar system knowing only their current parameters. The calculations of orbits was a sort of industry in the nineteenth and early twentieth centuries, requiring patience and long tedious numerical manipulations by hand. The inventions of the differential analyzer (by Vannaver Bush in the 1930s) and later appearance of electronic computers (during the World War II) made possible in minutes or hours computations that had taken months or even years to complete.

ACTION, *VIS VIVA*, AND MOTION

In an age of "universalists," Gottfried Leibniz (1646–1717) stands out as one of the masters. Distinguished in almost all areas of philosophy, he was also the node of a network of savants through scientific societies and his journal, *Acta Eriuditorum*, and among his other activities independently developed the calculus based on integration rather than differentiation and created one of the first mechanical calculators. His fierce rivalry with Newton makes for a gossipy history as a clash of personalities and the social assignment of credit for discovery. But that's another story. The development of mechanics required many minds and many approaches. Once the *Principia* appeared its elaboration by the Continental mathematicians, inspired by the Cartesian program that Newton sought to replace, produced a very different foundation for forces and mechanics.

Leibniz's objected to Newton's invocation of "attraction" as a property of matter and, in general, to the "un-mechanical" approach of reasoning without firm causal foundations. The issue came to a head in the extended correspondence between

Leibniz and Samuel Clarke, the last editor and amanuensis of Newton. In a series of letters exchanged over a period of several years and finally published for public consumption, Clarke—acting as Newton's voice at the latter's direction—defended the methodology of the *Principia.* The core of his argument is that we can observe an effect and infer an action even without knowing the details of how either occurs. If we see a body moving in a curved trajectory, recall by the first law, this demands that something not only deviated but continues to deviate the mass. If we know that the "something" is a force, and we know somehow what the dependence of that force is on the masses and distances, we can determine the trajectory without knowing what the agent "is." Further investigation may yield a deeper understanding but the formalism suffices when the questions are about *what* will happen rather than *why*. The latter is left to speculation, or what Newton derided as "hypotheses." Leibniz distinguished between two "forces" in contrast to Newton. One, *vis muotor* (moving force), he argued, is responsible for the production of motion (the extension of the Aristotelian principle and the Newtonian construction). The other, the "living force" or *vis viva*, is an attribute of the body while in motion. The fall of a body yields work because the external force, gravity (not distinguished from weight) accelerates the mass and produces an impulse when the body strikes the ground. The quantity $\frac{1}{2}MV^2$ is the kinetic product of this work and this is the *vis viva*. While Leibniz did not include the factor of 1/2 (this was realized only later by Kelvin and Tait when writing their *Principles of Natural Philosophy* in the mid-nineteenth century), this doesn't matter for the general purposes of the argument. He posited that the quantity *vis viva* times *interval of time for the motion* is the *action* of the external force. This should be a minimum for the path chosen by the mass since a straight line is the shortest distance between two points, for a unidirectional constant acceleration coming from its potential to fall because it is lifted. This almost resembles natural place, and even more the principle of work from Jordanus, but it is a subtler concept.

Imagine we lift a weight to a height and then release it. At the ground, the impulse will be greater for increasing height, of course, but so will the velocity. No matter what height the body falls from, it will always reach a final speed that depends *only* on the height. We can, therefore, say that the quantity that measures the effect of the weight is not the impulse but the quantity that measures the motion, the *vis viva*. The difference between this and Galileo's conception is not trivial. While the impact of the body is certainly greater, this can be lessened by imagining the body to be compressible. On the other hand, for the *vis viva*, we are not imagining its effect on the surface at the end of its motion. Instead, we ask what it does during the fall, and how it accelerates. The longer the fall, the longer the acceleration, but the amount of *vis viva* gained will always precisely the effect of the *vis mortis*.

We can see this dimensionally, something that requires that we know that a force is the change in the momentum with time. The work, or potential for motion, is the product of force with displacement and thus varies as $(ML/t^2) \times L$. But the velocity has the dimensions L/t so the work is the produce of MV^2t, which is the action. The centrifugal force is Mv^2/L for displacement around a curve with length scale L so the work again has the same dimensions as the *vis viva*. Leibniz's hypothesis was that the action is always minimal along the trajectory for an accelerating body acting under the influence of a "dead" force (in the case of freefall, its weight). This is the first statement of the mechanical principle of *least action* to which we will return later in the book. This quantity was introduced into physical reasoning at an intermediate approach between Galilean and Newtonian dynamics when the distinction between weight and mass was still unclear. Leibniz chose, by happy coincidence, the one case where the distinction isn't necessary in classical physics. Since all bodies fall at the same acceleration regardless of their mass, for the purpose of defining action we can equate mass and weight. Second, the definition involves an integral, which means a sum, and leaving aside problems with the definition of the integral that would not be cleared up for several centuries, there is no need to think in terms of instantaneous quantities such as the speed at any moment. Finally, this is a quantity that has no direction; it is an attribute of the falling body along the trajectory, whatever the trajectory is, so we don't need to think about the reaction to an imposed force in any specific direction. It's a *total* quantity in the sense that it belongs to the moving body. Recall that Galileo, through his experiments with inclined planes, was able to separate the inertial component of the motion, the horizontal speed, from its vertical, accelerating part and thereby note that the vertical component alone determined the restitution of the body to its original height at rest, frictional and resistive effects notwithstanding. Leibniz took this a step further and allowed for the possibility that some of the *vis viva* would eventually transform into compression for elastic bodies.

Although the language is anachronistic, we can say that the amount of energy is conserved by a moving body, whatever its form, as is the impulse or *momentum*. This permits us to treat collisions as well as motion under constant forces. For an impact not only accelerates a body from rest and imparts a momentum, it also gives the body a *vis viva* that is transferred from the collider to the target. We can then say, as Leibniz did, that the total momentum *and* the total *vis viva* remain with the movers throughout time unless something else brings the bodies to rest. One, momentum, acts along the direction of the collision. The other is independent of direction. This is consistent also with Descartes' concept that the quantity of motion in the universe is constant, just redistributed among bodies as they interact. The new approach opened a vast territory to mechanicians who now could apply a small number of physical principles to a huge range of problems without feeling constrained to explain the origins of the forces involved. Instead of hypothesizing a mechanism and seeking an effect, experiments and observations were now the tools for understanding the general properties of matter by rigorously reasoning from effects back to causes.

ELASTICITY AND FLUIDITY

The new machinery of the calculus and the new laws of mechanics were applied to more than just theoretical questions. They had a very practical consequence, engineering. While Galileo had begun the reform of statics, now problems of stability and response to changing conditions could be addressed. During the eighteenth century, the interest in dynamical reactions of objects to distortions and loading increased as technological developments required better understanding of the strength and behavior of materials. As projects grew in scope and complexity,

Figure 4.5: Leonhard Euler. Image copyright History of Science Collections, University of Oklahoma Libraries.

and with the definition of force allowing quantification of the conditions governing machines and structures, similarities were seen between what had previously appeared to be very different problems.

Leonard Euler (1707–1783) was among the first to apply the new perspective to the dynamics of bars and plates, leading to a general theory of material elasticity. Let's look at how it was accomplished. A bar can be viewed as a collection of strings (imagine the analogy between a coaxial cable and a cylindrical bar, for example). Then if one end is clamped and the other end is loaded with a weight, the bar bends. When Galileo treated this problem, you know he considered only the breaking point of the load and the scaling laws for the strength of the beam. On the other hand, we might want to know what vibrations would be possible for this object, depending on how and where it is clamped.

Elastic media provided a superb challenge for Euler and his contemporaries for applying force concepts beyond collisions and trajectories. It also required a modification of the concept of inertia. For a point, this is the total mass. But for a string, or a membrane, or a prism or beam, each tiny piece of the body is interconnected with the others and the calculus becomes indispensable. The matter is distributed continuously with some density and the force is also distributed over the volume. A string or spring is an ideal example. Hooke had considered only equilibrium. Now, having the relation between force applied and the reaction produced, it was possible to ask questions about what happens if this equilibrium is disturbed. Pulling the string induces a distributed reaction, the tension, and the strain depends on position. Its density may also depend on position, the mass per unit length may not be constant (depending on how the string was made). If you release or reduce the applied force the string moves to, and then continues past, the equilibrium. If it compresses, there's a contrary reaction that produces an extension, on expansion the sign of this reaction reverses, and the resulting motion is cyclic. For small deformations, any elastic material behaves according to Hooke's law and the tension is linearly proportional to the strain with the constant of proportionality depending on the properties of the material, the Hooke constant. Since the reaction is a force, there is a characteristic timescale for the oscillation that depends only on the length of the string (its total mass) and the strength of the reaction. In other words, we can *compute* the frequency of the vibration and thus obtain the string's elastic properties by looking only at the oscillation. Then we can extend this to two or three dimensional things, such as sheets and beams, by imagining them to be composed of networks of masses with string-like interconnectedness.

In both cases, we have a *harmonic oscillator*, one that has a fixed frequency that is independent of the amplitude of the vibration and only depends on the size of the equilibrium system (in this case, the elastic constant, mass or moment of

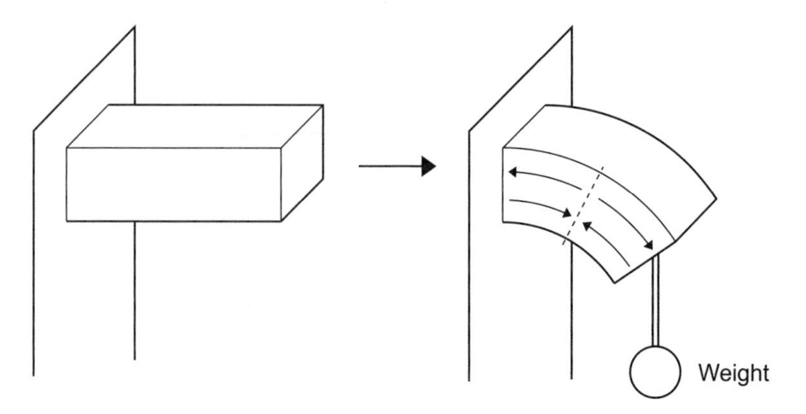

Figure 4.6: The bending of a beam showing the differential strain and stress described by the two-dimensional application of Hooke's law of elasticity. If the weight is released, the beam will oscillate above and below the horizontal.

inertia around the point of oscillation, and total length). The most familiar form of Newton's second law, $F = ma$, is due to Euler and serves when the mass (inertia) is constant for the mover. Then for a spring, the one dimensional change in length produces a reaction of equal magnitude and in the opposite direction depending only on the extendibility of the material—the constant of proportionality found by Hooke, K. When there are no other forces acting other than the spring itself, the motion is periodic and independent of the amount of stretching. The frequency depends only on the elasticity and the inertia of the spring plus the attached mass so a stiff spring has a shorter period for any mass, and a heavy load oscillates with longer period. We can generalize this to the case of a continuous mass loading, that the spring is moving under its own weight for instance, and then we have the same case as a pendulum with a moment of inertia. If the spring hangs freely, without loading, its mass produces a variable stretching that depends linearly on distance but if disturbed the oscillation has this characteristic frequency.

We can again make an appeal to dimensional analysis. We have the three laws of motion, and the basic units of mass, length, and time. Call these M, L, and t. In these units, the change in a length dimensionally L. Thus Hooke's law can be recast, dimensionally, as $ML/t^2 \sim KL$, so the units of the constant K are M/t^2. But this is the same as a mass times the square of a frequency since the unit of frequency (now called Hertz or Hz) is $\omega \sim 1/t$. Then the frequency of oscillation of a string, ω, for which the Hooke law holds precisely in the small deformation limit, scales as $\omega \sim \surd(K/M)$ for a string of mass M and the dispersion relation—the connection between the frequency and the wavelength of the vibrator—becomes $\omega^2 \sim (K/M)L^{-2}$. Thus the higher the tension, the higher the frequency, and the higher the mass the lower the frequency and this gives a wave that moves along the string. To extend this to a beam requires changing to a stress but then the moment of inertia of the beam enters and the frequency scales as $(Y/I)^{1/2}$ where we take Y, the Young modulus, to extend Hooke's law to the finite width of the beam. Now since the dimensions of $Y/I \sim L^{-4}$ we have for the dispersion relation $\omega^2 \sim (Y/I)L^{-4}$. This is different from our previous result for the string because we need to include the mass per unit area and the force per unit area (hence the

L^{-4} dependence). A beam has a variable radius of curvature that depends on its thickness (it's a case of a differential force, or shear).

The important step Euler took was to extend this to the more general case of displacement in two or three dimensions. A string, is a springy line. Plucking it, the force depends on the change in the length of the string because the displacements are directed along the perpendicular direction. For a symmetric displacement, Euler could write the force in terms of the difference in the slopes of the cord at each point along its length, x, and then take this as a second derivative with respect to that coordinate which yields the same dispersion relation we found using dimensions alone. This also provides a very important insight into the meaning of inertia, one almost anticipated by Buridan in his *Questions on the Physics*. Recall his statement that the string is a continually oscillating body because it is constantly re-traversing its equilibrium once it's gained an *impetus* by a violent motion. I say *almost* because the foreshadowing is really illusory. Buridan had no idea of inertia and that's why the string passes its stable point: once the unbalanced force accelerates the distorted string toward the midpoint, when the force vanishes the motion doesn't stop. How different this is from the Aristotelian force law! You see here why the idea of mass was so critical to the final formulation of a proper dynamical principle. With the requirement that motion requires unbalanced forces, it is obvious the string should come suddenly to a dead stop at the mid-place. But when the acceleration vanishes, the momentum doesn't and the passage of the string through the midplane reverses the sign of the reaction. Thus the string will again distend as much as it had originally, ignoring dissipation, and finally will reverse motion since again the forces are maximally unbalanced at the extreme of the oscillation.

For a linear displacement, Hooke's law suffices to describe the reaction force. But there are other ways to deform an elastic solid and this requires a more complicated theoretical machinery. Although for a wound spring it would seem you need the notion of a twist, this wasn't included. Instead, a spring—for instance, in a watch—was assumed to be locally extended and the force was caused only by the expansion or contraction of the length. But this isn't the same thing as the twist of a filament. Imagine a wire that is pinned at its two ends. Applying a torque at only one point produces a helix whose tilt angle changes as the wire is twisted further. In the limit of small deformation, the reaction is described using only the angle $\delta\phi$ and now with the moment of inertia instead of the mass, the oscillation frequency is similar $\omega^2 = E/I$. The coefficient is the same as for a beam, and emphasizes why the study of strength of materials is so connected with the Young modulus. Shear and twist are really the same thing, they're both nonaxial, torsional, distortions. This would later provide the necessary tool for the measurement of the force law for charges and the gravitational constant (by Coulomb and Cavendish, see Chapter 5). As Galileo had discussed in the *Two New Sciences*, there is also a non-elastic limit for materials, a point beyond which the body remains permanently deformed or even breaks. Euler's treatment permitted the introduction of higher order terms in the force law, which produce anharmonic motions, and permitted the calculation of the critical limits to the distortions.

Euler and Lagrange were also, at the time, engaged in fundamental problems of mechanics so it was natural that their thinking would combine their various interests when it looked as if the same techniques could be applied. So when thinking about membranes and plates, they were also working on fields and fluids and the inevitable cross fertilization was very fruitful. Particle dynamics didn't require a continuous medium but such questions of flexible response did. The extension this treatment to the distortion of clamped beams in equilibrium to beams in oscillation was not as simple but became the foundation of quantitative engineering mechanics and was again posed in the context of least action. Imagine the beam again to be a collection of fibers each of which is stretched by a length $l\delta\theta$. The bending of the beam produces a differential stretching that does work through elasticity. The Young modulus, E, and the moment of inertia I are required to account for the finite cross section. The bending is differential so there is a second order effect that is absent for a string or thin membrane but there for beams and plates: the differential stretching of the medium across the area transverse to the axis of the beam or plate produces a second degree rather than first degree dependence.

This period also saw the development of a new science, acoustics, based on the ideas derived from statics and small perturbations of elastic bodies. Newton had treated the speed of waves in continuous media, including the compression of air and the motion of waves in channels, but the interaction between vibrating sources and their surrounding media is a much more complex problem. It's one of boundary conditions. For instance, if a plate is in equilibrium under its own weight, statics tells you how it deforms. But if a rigid plate is set into vibration, for instance if you hear a gong, the vibration of the plate is being coupled to the surrounding air, producing a wave that carries energy outward from the plate. Once again the problem becomes partly geometric, the symmetries of the plate determine its vibrational modes. A membrane is the extension of a string to two dimensions since the moment of inertia doesn't matter. Hence we can go to two dimensions with the combined motions being only in the direction perpendicular to the plane. This acts like a piston on the surrounding medium which has its own elastic properties (this also worked for fluids in general). The local oscillation in one dimension becomes a forcing in three that begins at the surface and extends through the whole volume. This is a wave and the solution of the equation that describes the motion of the driving body, the boundary conditions, and the symmetry of the medium and the surface, since we could also write this in cylindrical or spherical coordinates depending on the shape of the oscillator. This was eventually demonstrated experimentally by E. F. F. Chladni (1758–1827) who studied the vibration of plates and membranes driven by stroking the edges in different places with a violin bow. He rendered the vibrations visible with a remarkably simple technique: knowing that the surface would consist of nodes and extrema of the oscillations, he reasoned that the nodal pattern would be stable to a steady stroking. Then sprinkling a fine powder or sand on the surface would lead to accumulation of the grains at the stationary points of the surface, avoiding those parts that are in constant motion. The method showed how to study the harmonic motions of any surfaces, even those with shapes far more complicated than could be computed.[10] There are engineering implications of this elastic force law since a structure may have resonances, depending on its elastic and inertial properties, that can cause catastrophic failure. This is because of the *natural* frequencies of the structure. If the driver happens to coincide with some rational multiple of these frequencies, unbounded amplification can occur limited by the point of inelastic response of the vibrator. An important application came after the invention of the steam locomotive and the expansion of railroads, the vibrations of tressel structures and arches became an important area for practical mechanics within a century of these first applications of continuum mechanics.

But there's an even more dramatic application. Imagine you're sitting in a quiet room and suddenly the floor begins to vibrate. The motion may last for only a moment but it is an unmistakable experience. The building may sway, lamps may swing, dishes and pictures may fall, and the walls may crack. This is what it's like to be in an earthquake, something that is familiar in many places on Earth. Earthquakes were variously explained by invoking only local effects without considering the response of the planet as a whole. Quite naturally, the explanations were of the same sort as for volcanic explosions: changes in the sub-terranian environment that caused upheavals of fluid and expulsion of gas. The local motion of the ground was in response to this. The seismograph, at least the ability to measure the moment and direction of ground motion, appears to have been invented in China in the second century by Chang Heng. Using a simple dropped pellet, the device registered the direction of the ground motion but gave little indication of its amplitude. This is just the opposite case to Chladni's visualization method, the driving of a mass-loaded spring by the ground motion leads to a visualization on a chart of the oscillation of the spring. There isn't just one frequency, the motion may have a very complicated spectrum, but elasticity theory can treat both the local and global motion of the Earth.

FRICTION

Aristotle's law required a driving force to maintain motion against resistance without identifying what either force is. In freefall, this was presumed to be the air since it has weight. But for motion along a surface this was less obvious. Think of a cube being dragged across a table. The friction opposes the motion but it's along only one face of the body. It must be due to some sort of shearing because of the relative motions but what? And when the table is tilted, the cube does not move. This would have meant, before Newton, the absence of any force. After him, it meant the forces were in balance. But how?

In equilibrium, the reaction of a surface to the perpendicular component of weight is the *normal force*. For hydrostatic equilibrium, this is the same as for a horizontal table. But for the inclined plane, the motion is impeded by a component of the weight parallel to the surface while the normal component remains in balance as a constraint on the motion. The parallel component is the acceleration of gravity, which comes from outside, reduced by the friction. With such immense practical importance this mechanical effect provoked many studies. The most systematic was performed by Charles Augustin Coulomb (1736–1806) who, at the time, was engaged in various engineering studies stemming from his military duties. Coulomb's law states that the frictional component, whether moving or static, depends only on the normal component of the mass and is independent of its velocity or area. It is therefore a constant fraction of the normal component of the force whose value depends only on the properties of the surface with which the mass is in contact. For an inclined plane, a block will remain static until the component of the weight parallel to the plane exceeds the friction. There is, consequently, a critical angle that is independent of the mass given by the *angle of repose*, $W \sin\theta_c = \mu W \cos\theta_c$ from which the coefficient μ can be determined. Several theoretical models were proposed, even incorporating various ideas about the distance dependence of a force. When in contact, the irregularities of the surface suffice to explain why a body cannot move freely and what can hold it in place.

Frictional forces have a special significance. Unlike any that can be obtained from a field, they are not conservative. That is, they cannot be obtained from a potential function (see the next chapter). Since the moving body eventually comes to rest, it would seem a natural consequence that such forces are dissipative. This is a fundamental limit to the efficiency of any real mechanical device. But this wasn't at all obvious at first and remained obscure until work by John Lesley (1766–1832) and Humphrey Davy (1778–1829). And because friction is a damping action, a loss of *vis viva*, it must be a different kind of force than gravitation. Herman von Helmholtz (1821–1894) realized how to include frictional effects in a more extended framework. His fundamental paper of 1847 was entitled as *On the Conservation of Forces* but in it, we find a more complete principle, conservation of energy, that replaces *vis viva* and its distinction from *vis mortis*. Frictional dissipation was essential for this generalization. Motion resulting from a potential can be lost but its "quantity," the energy, is redistributed between bulk (center of mass) motion, possible mass losses (for instance, if something breaks), and heat (internal energy). Deformations during collisions, for instance compressions, stress the body; if this is inelastic some of the kinetic energy is lost but the sum of all possible forms remains the same. Friction of any kind transforms ordered motion into something else, heat and possibly changes in the state of the material. We will discuss this further in the context of thermodynamics. Force, in Helmholtz's sense, was not just acceleration butt included action. It also encompassed changes in the state of the body, for instance melting.

Viscosity: The Internal Friction of Fluids

Daniel Bernoulli and Euler had been able to treat the motion of ideal fluids for which streamlines maintain constant total pressure and move without friction relative to each other, described in a memorable phrase of Richard Feynman as "dry water." There is a partitioning of the bulk motion that transports the medium of density ρ in a net direction with speed V and the internal pressure. The theorem, due to Bernoulli in his *Hydrodynamica* (1738), states that in the flow the sum of the dynamical pressure, P, or the momentum flux, and the internal pressure remains constant along the flow, $P + \frac{1}{2}\rho V^2 =$ constant; here ρ is the mass density and V is the mean velocity. This actually *defines* a streamline. Pressure differences within the medium drive the flow, along with external forces such as gravity. These produce continuous changes in speed along the streamline as well as in time. Euler's equation of motion describes the hydraulic motion requiring only boundary conditions to define the specific problem. The pressure acts outward as well as along the flow direction. For instance, in a pipe a change in the pressure accelerates the fluid but also acts on the walls so the elastic problem of the expansion of the container becomes coupled to the problem of describing the motion of the fluid contained.

The difficulty is this is very incomplete. It ignores the drag at the walls because the fluid is assumed to be frictionless. Real world observations, for instance canals and rivers, show that this is too idealized for many applied problems. For example, in channel flow the velocity at the boundaries, perpendicular to the wall, must vanish and if there is no slip there, so the flow is also stationary at the wall, then a gradient must develop in the pressure that points into the flow. The velocity at the center of the channel is largest, the velocity profile is parabolic relative to the center of the channel. A nonsymmetric flow, for instance over a flattened sphere, generates a differential pressure gradient in the fluid and produces lift. If you have a curved surface, the flow accelerates because of the change in direction along the boundary. We imagine the curved part to be on top. Along a straight edge it is inertial, let's take that at the bottom. The pressure is therefore different on the two surfaces and a gradient points upward. This is the principle of the *airfoil* and the basis of the theoretical treatment of flight. But the Euler equation and Bernoulli's principle are only the first approximation to the real mechanics of flight. The treatment has to be extended to include the effects of a distributed friction, the viscosity, between the moving body and the surrounding medium and within the fluid itself. The momentum transfer is to a bulk, deformable medium that occurs on many scales depending on the shear.

You may wonder why this long digression in a discussion of forces? For a continuous medium, this viscosity depends on the material properties of the medium. The definition of a fluid is that it is a continuous medium that doesn't resist shear, or differential distortion. In contrast, from the bending of solids, the Young modulus or Hooke constant measure the resistance to deformation. Even for an incompressible medium, the shear is the difference in the momentum between two parts of the fluid, or between the fluid and any bounding surface. The first solution for a fluid was by proposed in 1845 by George G. Stokes (1819–1903), although a similar result was obtained by the French engineer Navier somewhat earlier using a very different picture of the interactions between parts of the medium. The importance is that the friction depends on shear, the differential motion of the fluid. This dissipation, or "'loss of head" (impulse), was the essential ingredient missing from hydraulic calculations of fluid transport. There are several sources for viscosity, and these are important for extending our idea of a force. Coulomb's picture of friction was a contact force between solids. But in a fluid this arises because of shear and momentum transfer at the microlevel. Macroscopic momentum transport also occurs because of turbulence, a state of the fluid first encountered in engineering applications of flows through pipes. Again, the details would require far more space than we have available. It is enough to say the forces involves are friction versus the inertia of the fluid, the ratio of which form the *Reynolds number* (a scaling discovered by Osborne Reynolds in the last third of the nineteenth century), $V L/\nu$ (where ν is the coefficient of kinematic—internal microscopic, or kinematic—viscosity and V and L are characteristic scales for velocity and length within the medium independent of the geometry and boundary conditions.

ROTATIONAL MOTION AND FRAME OF REFERENCE

After the *Principia* something was still missing in the mechanical picture: a precise specification of how to treat space and types of relative motion. It's almost as if Newton had "thrown out the baby with the bathwater" by banishing any discussion of *centrifugal* acceleration from mechanics. For the central field problem with which he was most concerned for gravity this was sensible. But it forced a unique choice of reference frame, that of a stationary observer who would understand what motion is or is not inertial. The relativity of the motion, and its independence of frame of reference, was ignored and including it provided a new understanding of the meaning of a force.

I'll begin this discussion by noting an odd historical fact: Newton never wrote the iconographic form of the second law of motion in the form that now appears in every introductory physics textbook, $F = ma$. As I'd mentioned earlier, this simple expression was another of Euler's many contributions to mechanics. It may come as a surprise that to achieve this compact statement, which is actually not the same as Newton's, required almost half a century. But stated this way, force takes on a new significance. For Newton, all forces are real, existing things that require sources. For Euler, they are—as the second law states—the agents that produce accelerations, for Newton, forces are measured by accelerations. But the two are interchangeable and what you call a "real" force depends on your point of reference. There *is* a centrifugal *acceleration* that with a simple change of wording becomes a *force* if you happen to be in a rotating frame and, perhaps, don't realize it. Dynamics in a rotating frame appears different, as we saw from the Newtonian bucket problem, because the matter is continually accelerated. Thus, we must include the motion of the "container" as well as motion within the container when describing any accelerations. A simpler form of relative motion was already discussed by Galileo, rectilinear motion at constant speed. This *inertial frame* is one that leaves the equations of motion invariant, since the center of mass now becomes a reference point. In the equations of motion, we can choose which frame is appropriate, the two should give the same answer for any measurable quantity. We now refer to these as the Eulerian (fixed in a stationary frame outside the motion) and Lagrangian (relative coordinates with respect to the center of mass or mean motion).

To understand this, remember that in a central force problem only the radial component of the motion is accelerated. That's why for a two body orbit the angular momentum is strictly constant unless there is a torque, for instance if the central body is rotating and extended or deformable by tidal accelerations. In a rotating frame, any radial motion requires an increase in the angular momentum to stay in corotation so as you move inward or outward from the axis you experience *what seems to be a force* that wouldn't be there if you were in a stationary system. On a disk, this is very apparent, think of trying to walk from the center to the edge of a carrousel: you feel an apparent deceleration *in the angular direction* making it very hard to proceed radially outward. On a rotating sphere, when you are walking along a meridian you are actually walking radially relative to the rotation axis. You are moving toward the axis (displacing from the equator) or away (going in the opposite sense) although you are maintaining a constant distance from the center. The same is true when you move vertically (upward motion increases the axial distance in either hemisphere). Thus the motion, seen in the rotating frame, seems to have a twist—or *helicity*—and an acceleration always a right angles to the direction of motion. The sense of this is toward the right in the northern hemisphere of the Earth relative to the direction of motion (and opposite in the southern hemisphere). This is the Coriolis force, named for Gaspard Gustave de Coriolis (1792–1843), and it is a dynamical consequence of the kinematics of a rotating system.

To be more precise, the force depends on the velocity, not merely the displacement. The faster the rate of displacement, the greater the rate of change of the angular momentum, which is an angular acceleration or torque. But the acceleration is in the tangential velocity and thus seems to be due to a tangential force. The practical example is any motion on the Earth, toward or away from the equator. The deflection depends on the rotational frequency of the surface, the rotation frequency of the Earth, and the latitude, which determines the angle between the local vertical and the direction of the rotation axis in space. The existence of this force was dramatically demonstrated in 1851 by J. B. Léon Foucault in (1819–1868) with a very long (11 meters) pendulum suspended from ceiling of the meridian hall of the Paris Observatory and later, in a public demonstration, from the dome of the Pantheon in Paris. The Coriolis acceleration also dominates the motions of the atmosphere, especially local systems of high and low pressure and explained the laws for the directions of winds.[11]

Thus, relative to an observer at rest—that is, not in the rotating frame—there is always an angular motion associated with any change in position. This was first discussed by Euler. Even an observer who isn't moving within the rotating system, for instance someone sitting on a carrousel, is still being carried around at the angular velocity, ω, relative to an external observer, who is also at rest but in a truly stationary frame. This is not invertible. Although we can change the sense of the rotation by making the whole universe rotate in the opposite direction from purely kinematic point of view, the essential difference is that rotation, unlike rectilinear motion, is never inertial: the motion takes place in at least two dimensions that change in time and is therefore always accelerated.

This was the basis of Ernst Mach's (1838–1916) discussion of force. It gets to the core of what inertia is all about, that there is some sense in which relative motion is *not* symmetric. Newtonian cosmology assumed an absolute reference frame and an absolute time. But somehow the moving system "knows" that it isn't inertial because forces appear in one frame of reference that another doesn't detect, seeing instead a different motion. The Coriolis and centrifugal forces in rotating frames are the principal examples. Mach's solution was that even if all motions are relative, on the largest scale there is a semi-stationary distribution of the mass of the universe as a whole with respect to which all local motions are referred. This became known, much later, as "Mach's principle." It is still an open question whether this is the origin of inertia, as we will see when discussing the general theory of relativity and gravitation in the twentieth century later in this book.

A way of seeing this uses another device invented Foucault in 1852, the gyroscope. Although it is now a children's curiosity, it is a profound example of how non-intuitive the effects are of rotation and angular momentum conservation. Euler showed that a non-symmetric mass will tumble around a principal axis of inertia so its apparent rotation axis seems to shift over the surface of the body on a period that depends on the distortion of the body from a spherical shape and on the rotation period. This freebody precession, in which the angular momentum remains fixed in space but the instantaneous axis of rotation shifts around, was finally detected for the Earth by Seth Chandler (1846–1913) using the motion of the north celestial pole. Remember, an observation by an observer on the surface is made in a rotating frame. The period is about 14 months for a distortion of about 0.3% and a rotation period of one day with an effective displacement of the order of meters. The precession of the equinoxes is a much greater effect, caused by the distortion and the gravitational attraction of the Sun and Moon on the tipped, spinning Earth. Foucault's gyroscope consisted of heavy disk that spun rapidly around an axis that was attached to an outer support ring that could be either suspended on a cord or balanced on a fulcrum, much as it is nowadays. When tilted the disk is torqued like a lever and tends to right itself. But in so doing the angular momentum must change direction which produces a motion around the vertical, the direction of the gravitational acceleration. The disk is far more flattened than the Earth but the principle is the same. Foucault realized that the device provides a reference for the direction of gravitation and it was soon employed as a stabilizer for ships and, eventually, for satellites and missiles.

NOTES

1. The texts are taken from the translation by Cohen and Whitman (Newton 1999). This is the best introduction in the English language to the history, context, and contents of Newton's monumental work and I urge you to read it.

2. In effect, he showed that Galileo's treatment of freefall is the limiting case when the distance traversed is small with respect to that from the center of the force so the acceleration is approximately constant (see the "Universality" section).

3. This is the line passing through the foci of the ellipse that connects the points of closest and farthest approach to the central mass.

4. In many textbooks in introductory physics it's popular to say the centrifugal acceleration is fictitious. From the viewpoint of an external observer that's certainly true and that's what Newton understood. But when you are reckoning the accelerations, if you're not moving inertially there is a force. It doesn't do to simply dismiss it, and this was one of the critical steps that led to the general theory of relativity and that we will meet in Einstein's principle of equivalence in a different form.

5. This is how Charles Darwin summed up his *On the Origin of Species*. It can equally be applied to Newton's *magnum opus* and perhaps in the same guise. Darwin used a similar principle to that employed in the *Principia*, explaining the phenomena of diversity of life in terms of universal mechanisms, the origin(s) of which were yet to be discovered.

6. In fact, this is still true. Although the pendulum has been replaced by bars and balanced oscillators driven at specific reference frequencies in modern gravitometers, the principle is essentially unchanged. A loaded pendulum, normal or inverted, stable or unstable, is used to measure the acceleration of gravity. An even more dramatic example is the arrangement of the gravitational wave interferometer Virgo that uses an inverted pendulum to serve as the test mass for detecting change in the path of the light.

7. I'm using here the Motte translation of 1727, prepared under Newton's guidance (that is, direction), since it is the one version that immediately connects with the intent of the *Principia* and was also read by his contemporaries who were less schooled in Latin.

8. Halley was the first to see the contradiction in this and although it departs from the main thread of our discussion of force, it is instructive to see how rich the Newtonian worldview was in new problems. If, Halley argued, the light from distant bodies diminishes similarly to their gravitational influence, then the it would seem that an infinite universe contradicts experience. Because the number of sources intercepted by a cone of fixed solid angle grows as r^3 while the intensity of any individual star diminishes as $1/r^2$, the summed intensity should increase linearly with increasing distance. Hence, the night sky should be infinitely bright, an unacceptable conclusion. Yet to imagine a world with an edge would have two equally unacceptable consequences. There would have to be a boundary, at finite distance, beyond which there would be only void. And the stability of the cosmos again becomes a problem. See Jaki (1973), Harrison (2000), and references therein for history and ultimate the resolution of what became known as *Olber's paradox*.

9. This was not the first "new planet," of course, Uranus had been discovered by chance by William Herschel in 1781. But there were other, so-called minor bodies, that had been discovered before Neptune was predicted. The first was Ceres, the largest of the asteroids, followed by Pallas and Vesta at the start of the nineteenth century. Their number steadily increased over the next two centuries.

10. This technique has been extended in the past few decades to other visualization tools, especially interference of lasers reflected from the surfaces, and even dynamical imaging is possible. But the basic idea has not changed since Chladni's first investigations.

11. A low pressure region produces locally radial motion as air moves from the surroundings into the low. This produces a vortical motion that is counterclockwise, or *cyclonic*, in the Northern hemisphere. It isn't true, by the way, that this is observable in bathtubs and sinks or even tornados; the helicity of these flow is imposed at their source, pipes and downdrafts in wind-sheared cold atmospheric layers, and is due to the conservation of circulation rather than the large scale Coriolis effect that dominates structures such as hurricanes.

5

FORCES AND FIELDS

> All physicists agree that the problem of physics consists in tracing the phenomena of nature back to the simple laws of mechanics.
>
> —Heinrich Hertz, *The Principles of Mechanics*

Even while the *Principia* was being read, digested, and debated, a complementary picture to Newton's was taking shape on the continent at the end of the seventeenth century where the Cartesian influence still held sway. The nagging difficulty was that there was an essential ingredient missing in the Newtonian program: while the effect of gravity could be described, both its source and transmission remained unexplained. Notwithstanding his metaphysical justifications for this lacuna, to say this bothered Newton's contemporaries is an understatement. It was the major stumbling block, in the context of the Mechanical Philosophy, that an action should be invoked without cause. Newton's own analogies served poorly to assuage the misgivings of his readers, especially Leibniz and the continental mathematicians. It seemed to call for some sort of action at a distance, an effect that almost miraculously issues from one body and affect another regardless of where it is located. To this point, all treatments of force had required some sort of contact. The Cartesian vortex, acting through a subtle fluid that pervades space, at least satisfied this requirement. But Newton had argued, in the extended polemic of the second book of *Principia*, that this would not suffice to produce the inverse square law. While this may seem to be the heady stuff of philosophical debate it was far more. It called for a fundamental change in how to reason physically about Nature.

The idea of a"field" as the origin of motive power was born in a sort of theoretical compromise and even with the antipathy to "attractions" and "influences" as words, the concept of field was very vague at first and was couched in language that resembled the Cartesian plenum. But it was, somehow, different. Instead of having a dynamical role in the creation of gravitational attraction, the field was something affected by masses to produce the desired effects. Spatial differences in the intensity of the field became the forces and a body was assumed to move continuously under their influence because the medium—whatever it might be—is continuous. This permitted the source of the action to reside in the massive bodies which, you'll recall, had been defined rather than deduced in Newton's program. Their influence on other masses, their attractions (the very concept that had been so philosophically perplexing), became the continuous changes in the density of this universal medium. An interesting symbiosis developed. Increasing familiarity with the properties of fluid and elastic substances produced useful analogies, and the foundational issues spurred further work along these lines.

This shift is vitally important for understanding all subsequent developments of physics. Dynamical questions were now to be posed in two steps. The first required postulating what kind of field was involved given its source (although still not its "nature") and then deriving the force through the its spatial dependences. Then, knowing how the force behaves, to compute the trajectory of a mass (or a collection of masses) thus acted upon. The basic problem was thus buried, not solved, and with this new-found freedom it was possible to make rapid progress. In effect, to know—or hypothesize—the field was to know the force. Or, conversely, to know the dynamics was to know the force and through the motion to determine the field. In this view, the physical scale at the phenomena occur doesn't matter. If forces are universal, although perhaps different in nature depending on the agent and circumstances, the same methodology can be applied.

THE ATOMIC CLUES

You'll recall that Galileo and his followers had been concerned with the nature of air and the vacuum. Experiments were possible after Robert Boyle's (1627–1691) invention of the air pump. This provided a means for artificially changing the pressure in sealed volumes under controlled conditions. With it, Boyle discovered the first law for the internal pressure of gases, that the reaction force increases on compression. You'll recall that liquid water, being an incompressible medium, produces this reaction depending only on its weight above any level. This property was essential in early arguments about continuous media. A gas can change its density so its weight changes the pressure also by compression. The atomistic picture was perfectly suited to explain this behavior, as Daniel Bernoulli did by assuming a gas to be composed of particles that are in constant motion and collide both among themselves and with the walls of the container. The argument is beautifully simple. If we assume all particles have an momentum (or impulse) p, and collide elastically, then when they hit a wall they bounce without changing the magnitude of this momentum. The kick, Δp, given to the wall is then of order $2p$ since they simply reverse direction (remember the third law of motion). There is another factor to consider, the mean angle of incidence, because the particles are assumed to arrive from any direction and thus only a small fraction deliver the kick along the surface normal. The number of such collisions in an interval of time depends on the rate of arrival of these particles at the wall. If N is the total number in the container of volume V of velocity v and mass m, the rate of arrival of the particles at the wall per unit area per unit time is Nv/V. Since the force is the time rate of change of the momentum, mv, in an interval of time is $N\Delta p/\Delta t$, and the pressure is the force per unit area, this gives the pressure law $PV = \text{constant}$. This also provided a scaling law for the internal pressure, $P = p^2 N/(mV)$ for a gas of identical particles. Furthermore, because the impulse—the kick on collision—is delivered perpendicular to the walls of the container this explained Pascal's law for pressure. The result provided at once a successful explanation for how to apply the new force concepts to materials and also strengthened the atomic viewpoint already implicitly contained in Newton's conception of both matter and light.

Newton had, in fact, published a possible extension of his conception of force to fundamental interactions and the atomic nature of matter but not in the *Principia*. That remained a work addressed to the picture on the large scale. However, he allowed himself more speculative freedom in his other work, the *Opticks*. Unlike his *magnum opus*, this was first written in English. The first edition was published in 1704. It summarized Newton's studies of light and optics but also included a small set of questions (*queries*) in a closing group of essays on the nature of light, its propagation, and the relation of optical phenomena to those he had developed in the *Principia*. The work quickly appeared in a Latin edition prepared by Samuel Clarke with Newton's supervision and containing a revised and augmented set of queries. With the second English edition in 1717 we have Newton's statement of his "hypotheses."

Recall Newton's absorbing interest in alchemy. For the final query, number 31 in the last edition, he pulled out all the stops, composing what can be read as a symphony. The text opens with the direct statement of the theme:

> Qu. 31: Have not the small particles of bodies certain powers, virtues, or forces, by which they act at a distance, not only upon the rays of light for reflecting, refracting, and inflecting them, but also upon one another for producing a great part of the phenomena of Nature? For it's well known that bodies act upon one another by the attraction of gravity, magnetism, and electricity; and these instances shew the tenor and course of Nature, and make it not improbable but that there may be more attractive powers than these.

There follows a long catalog of chemical phenomena that show how combinations require some sort of interaction that, unlike gravity, depends on the specific properties of matter. He then opens the second movement, inferring by inverse reasoning the probable properties of the hypothesized forces:

> The parts of all homogeneal hard bodies which fully touch one another stick together very strongly. And for explaining how this may be, some have invented hooked atoms, which is begging the question; and others tell us that bodies are glued together by rest (that is, by an occult quality, or rather by nothing); and others, that they stick together by conspiring motions (that is, by relative rest amongst themselves). I had rather infer from their cohesion that their particles attract one another by some force, which in immediate contact is exceedingly strong, at small distances performs the chemical operations above mentioned, and reaches not far from the particles with any sensible effect.

Returning again to specifics, he describes how the forces might build macroscopic bodies for which, in the end, gravity alone dominates. And then he generalize the interactions:

> And thus Nature will be very conformable to herself and very simple, performing all the great motions of the heavenly bodies by the attraction of gravity which intercedes these bodies, and almost all the small ones of their particles by some other attracting and repelling powers which intercede the particles. The *vis inertiae* is a passive principle by which bodies persist in their motion or rest, and resist as much as they are resisted. By this principle alone there would be no motion in the world. Some other principle is necessary for putting bodies into motion; and now that they are in motion, some other principle for conserving the motion.

Newton then discusses collisions, showing that by composition of momenta there may not be completely in agreement with the principle of conservation of momentum. This leads, at last, to a picture of atoms as fundamental units of which bodies are composed, ending with the affirmation that "therefore I scruple not to propose the principles of motion above mentioned, they being very general extent, and leave their causes to be found out." The text then ends with a deistic coda and the piece draws to a close, the final chords being the General Scholium.

FIELD AS FORCE

For Newton, forces remained impulsive, even in the limit of an almost infinite number of collisions. Something emanates from a mass that produces an influence on distant bodies and changes the motion of neighboring masses accordingly. Ignoring the problem of propagation, this construct sufficed to empty the world of its ether and vortices. Voltaire reported this in his *English Letters* noting that, in France, one finds a world full of fluids and vortices while in England it was empty. That's why modifying the Cartesian program was not so difficult once the Newtonian *principles* were included. It wasn't necessary to completely abandon the fluid as an explanation as long as it acted according to the laws of motion and obeyed their constraints. The advantage was that the source for gravitation might still be elusive, but its action could be more simply explained. We rename it, and call it a field, and imagine it as a continuous substance. Within the mechanical philosophy this required that spatial changes of this field produce the force. Conservative forces such as gravity, the gradients are always in the same direction as the displacements so the increase in motion comes from the work done to place the body in its starting position. In order to produce a change of state of motion, this something must somehow be unequally distributed among things in proportion to their mass. If this seems too vague for your taste, it is nonetheless remarkable that even without specifying what this field *is*, considerable progress was achieved by concentrating only on how it acts. Just as pressure causes motion when it unbalanced, so it may be with this field. Spatial gradients produce a net force that is only counteracted—the reaction—by the inertia of the body.

It seemed possible, within this conception, to take up the challenge of the Queries but few tried beyond lip service and change in vocabulary, when they weren't lamenting the "occult qualities" that seemed to be again entering physical thought. One who did was the Jesuit polymath Roger Boscovich (1711–1787). He assumed static forces that depend only on the proximity of atoms but that the fields by which they interact may have much shorter range than the inverse square law. "It will be found that everything depends on the composition of the forces with which these particles of matter act upon one another: and from these forces, as a matter of fact, all phenomena of Nature take their origin" (Theoria Philosophiae Naturalis 1758). An important hypothesis was the existence of an intrinsically repulsive core to the atom, one that prevented collapse of matter on close approach. This catastrophic state is unavoidable in Newton's cosmology for a gravitation-only universe. Even while not assigning a cause to this atomic level force, Boscovich could not only explain the different states of matter but also provided a framework for understanding chemical combinations. He considered the field to be the fundamental property of matter, whose spatial dependence could be very complicated, having alternating zones of attraction and repulsion, postulating a universal field whose particulars depend of the "type" of atom. Here he made a significant change with respect to Hellenistic and Roman atomism. His atoms are really points endowed with mass (inertia) and the field. Electrical and magnetic phenomena could have furnished close analogs but instead Boscovich concentrated on the form that would be required for the law. He also realized that there could be "trapped states" if the field reverses sign, points of equilibrium around which the body could oscillate, but did not explore this further and it remained only implicit in his construction.

GEORGE GREEN: THE ORIGIN OF POTENTIALS

Even with the investigations of Laplace, Legendre, Lagrange, and Poisson, by the second decade of the eighteenth century the concept of field remained rather vague. The analytic link was still lacking. That was provided by the English mathematician George Green (1793–1841). He remains a shadowy figure in the history of physics. Largely self-educated, far from the centers of the scientific activity, with a privately published single memoir he introduced the methods and framework for uniting the Newtonian and continental programs that, when finally fully appreciated, revolutionized the concept of force and changed the course of physics. Green didn't simply introduced the term and concept of the potential in his 1827 memoir *An Essay on the Application of Mathematical analysis to the Theories of Electricity and Magnetism.* He showed that *any* force, electrical or gravitational, can be found from the spatial variation of a *field of potential*, a function that depends only of position without any particular directionality in space.[1]

The force law for gravity was originally derived by Newton from orbital motion by assuming that locally—within the orbit—the centripetal force constrained the motion. The first derivation of a general mathematical expression for how to unite the field and its sources with the force it produces was achieved by Pierre-Simon de Laplace (1749–1827) at the end of the eighteenth century. He assumed a mass within a space that is the source of the field which must give only a net radial force. What lies outside of this mass was unimportant as long as its influence could extend to infinity. The shape of the body didn't need to be spherical, Newton had

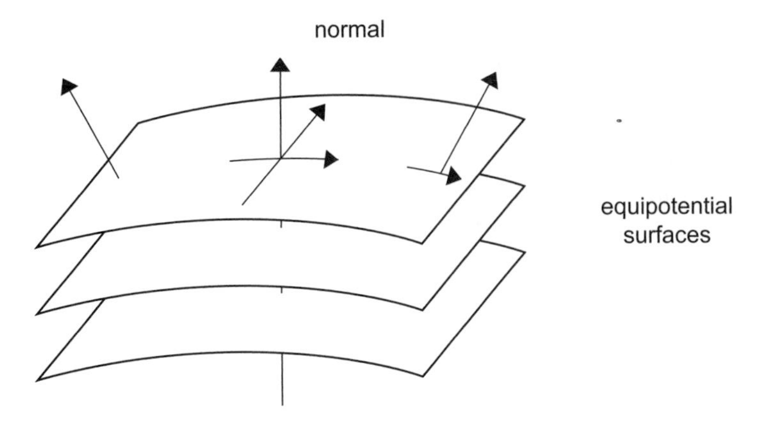

Figure 5.1: The construction of gradients through equipotential surfaces, an illustration of Green's idea of the potential.

already demonstrated that such a case is trivial in the sense that the resulting field is strictly radial. Laplace allowed for an arbitrary shape but remained outside the body so the force must, asymptotically with increasing distance, approach spherically symmetry. In this case, there can be no net divergence of the force: if we draw a sphere, or a spheroid, around the source of the field and extend its radius to infinity, the force must asymptotically vanish. The equation describing this was the starting point for the generalized concept of a field of force, stating that there is no net divergence over all space of the acceleration produced by a mass. It was straightforward, then, to include the effects of extended mass distributions. Poisson augmented the Laplace equation to treat a medium *with* sources, an extended mass distribution, was a relatively small alteration. Again, the boundary condition is that the mass is finite, if extended to infinity the force depends only on the amount internal mass. Each element of the mass attracts the test body so the net force is really a sum over all masses that are anywhere other than where we happen to be standing. Said differently, it is a *convolution*—literally, a folding together—of the density distribution in the sense that the force at a point within some density distribution is produced by the sum of all influences outside that point.

Now let me expand on this point for a moment. Laplace extended the force concept by introducing a *field equation*, the representation of the field as a continuous property of the *space* surrounding a mass. If we are outside all masses, then their *influence*, their collective force, has a very specific property. This is an essential feature of action at a distance, that the field once established fills space only depending on the distance between the affected bodies. Imagine a spherical mass in isolation, or at least far enough away from other bodies for their effect to be negligible. The force is then radial, directed toward the center of mass. The source of the field can then be treated as a point and we take the rate of divergence of the influence through successively larger spheres circumscribing the source. Saying that this divergence vanishes is the same as saying that whatever "effect" passes through a sphere of radius r will pass through a sphere of radius $r' > r$ and also $r' < r$. The same is true for light, the flux from a constant source is conserved so the amount of energy passing through any element of a circumscribed surface decreases as the inverse square of the distance; the total amount of light passing through the whole surface remains constant. Newton had hinted at this analogy in the *Opticks*, Laplace showed that the equation for the field then gives a general solution for any coordinate system and permits the calculation of the force, through an auxiliary function, by saying that if the force is the gradient of some functional representation—the field—then the divergence of that gradient vanishes. As I've mentioned, Simèon-Denis Poisson (1781–1840) extended this to the interior of a distribution of arbitrary density, demonstrating that the divergence of the force depends on the spatial distribution of the density of matter. For a point source this reduces to the Laplace equation. For electrical media, Gauss derived the same equation using charge instead of mass, that the electric field passing through any completely encompassing surface is normal to the surface and is only due to the charge contained herein. The mathematical generalization of this result to any arbitrary scalar quantity is now called *Gauss' theorem*.[2] We know from the introductory remarks to the *Essay* that Green had read Laplace's monumental *Mecanique Celeste* —at least some of the formal parts dealing with the representation of functions in spherical coordinates, Fourier's *Theorie analytiques del la chaleur*, the relevant papers by Poisson (but perhaps only in translation in the *Quarterly Journal of Science*), Biot's book *Traite de physique* from 1824—but was essentially self-taught and isolated from the centers of English and Continental mathematical physics. The problem Green was examining, how to find from the electric field the distribution of the internal charges, had already been confronted nearly a half century before in a completely different setting, that of gravity. Green defined the potential as a function, a scalar whose gradients produce forces but which is not itself measurable. His is the first use of the term although the concept was already built into *vis viva*. This step might be described as the bridge between *dead* and *live* forces, the field comes from the properties of distributed matter but the action of the distribution produces motion.

This potential function wasn't merely a *technical* modification of mechanics. It provided a way to explain and visualize forces. Now it was possible to use inertia again geometrically, not only for motion but also for statics. One could picture of field without asking about action at a distance, without emission or absorption of the agent responsible for the forces. It meant only imagining a landscape in which, for instance, a body placed at some point would be stable or unstable around its equilibrium. Instead of thinking of lines—trajectories—and directions of influences, it allows you to imagine surfaces and gradients. The mathematical connection with geometry, and in particular with *differential geometry*, was a decisive step. It's possible to say, for instance, that the potential *is* the field, that the forces are therefore produced by changes in the landscape in space and, eventually, in spacetime. For instance, if there is little change in the field at any place, or if you take a region small enough (or far enough away from a central source for the field, the local change in the direction of the surface normal gradients is as small as you want and therefore the force appears both vertical and the potential surfaces flat. How like the problem of mapping in geography. If you take a region sufficiently small in radius, the world is flat. As the view extends farther, toward the horizon, the surface may appear curved. That is, if the view is clear enough and that is always true in our case.[3]

Geodesy and the Measurement of the Density of the Earth

But to understand Green's revolution, we should step back for a moment. Geodesy was the child of the eighteenth century, a part of the systematic mapping of the Earth's surface. If the previous two centuries were expansionist, the 1700s were driven by a need for precision to maintain the products of that expansion. European colonies were spread across the globe and the seas were the open thoroughfare to pass between them efficiently. But it was necessary, on land and by sea, to have accurate maps. This required, in turn, precise knowledge of latitude and longitude and of the shape of the planet. For instance, in siting a meridian line—even with an accurate clock—it is necessary to know where the vertical is and also the altitude relative to a fixed reference frame. This was provided by gravitation. A model for the Earth was also required for altitude measurements since the swing of a standard pendulum (a "seconds pendulum") could provide the reference for time but required knowing the distance from the center of the planet. During the French expedition to Peru, under the auspices of the Academie des Sciences, to measure the length of the meridian arc for geodetic reference, Pierre Bouguer (1698–1758) developed one of the standard corrections for terrain gravity. He took account of both what is now called the "free air" correction, which is an adjustment for the altitude of the observing station relative to the reference spheroid, and for the mass of the underlying material, now called the "Bouguer correction." In effect, these are corrections not to the acceleration but to the gradient of the acceleration and require a knowledge of the geology of the underlying material or at least an estimate of its mean density.

Figure 5.2: In the background of this portrait of Nevil Maskelyne, notice the mountain that was the object of his study of the gravitational attraction of the Earth. Image copyright History of Science Collections, University of Oklahoma Libraries.

A fundamental test of Newtonian gravitation, albeit an indirect one outside the controlled space of the laboratory, was Nevil Maskelyne's (1732–1811) measurement of the deviation of a plumb line by the presence of a nearby mountain. It became one of the first geophysical measurements, the mass of the mountain was determined by the deviation and this was then, along with geodetic sitings of the structure, to obtain the contrast between the density of the surface and that of the bulk of the Earth. The result, that the density contrast was nearly a factor of two between the superficial and deep matter, might seem a mere curiosity. Yet this was a starting point for modeling a part of the Earth inaccessible to direct measurement and was important not only for the initial technical problem but even had cosmic significance. With this determination of the density *profile*, albeit only two extreme values, it was clear that the planet is more centrally condensed than the mean estimate based on the acceleration of gravity and the radius of the planet. This, in turn, affected the calculation of the lunar orbit and of the tidal effect. It was the beginning, as well, of the investigation that culminated in Henry Cavendish's (1731–1810) measurement of the density contrast of the Earth using the laboratory determination of the proportionality constant for gravitation, the universal value of which still lacked precision at the end of the century. The Cavendish experiment of 1798, as it is now universally called, used a torsion balance, a technique perfected for electrostatic measurements by Coulomb in 1784. The principle is brilliantly simple, even more so for masses than charges because the relative magnitude of two masses was far more easily determined than the relative quantity of electrification. Two test masses were attached to a suspended beam which was maintained in a thermally and pressure controlled case. A thin wire was attached to a micrometer and two masses, suspended from a second arm, were swung from rest to a fixed angular displacement. The test masses, thus accelerated, displaced and oscillated for hours, and the deflection was measured by a small mirror mounted on the wire to reflect a light beam to a wall where the deflection could be more easily observed and recorded. Although Cavendish intended this to be a measurement of the density of the Earth, its main result was the precise value for G, the strength of the gravitational attraction, so the proportional statement of the gravitational force law between two masses, M_1 and M_2 separated by a distance r could be replaced by a precise dimensioned statement:

$$F = -\frac{GM_1M_2}{r^2}$$

The measurement was repeated, with increasing accuracy, during the 1800s, summarized by C. V. Boys (who asserted that the principal result of the Cavendish experiment was the value of the gravitational constant), and more refined measurements, now using different techniques, continue through this century.

The Potential Function

Now, after this interlude, we return to George Green. The *Essay* was greeted with a deafening silence, at least at first. Few copies were printed, and surely many that were receive by even the subscribers went unread. Green's time in Cambridge was longer as a student than as a Fellow, a post he held for less than two years until he returned to Nottingham in 1841 where he died (the cause is unknown) a matter of months later. Copies of the work were also sent to such noted scientists as John Herschel (who had been one of the subscribers) and Jacobi, but it is doubtful they—or anyone else—initially appreciated the extent of the change of method and power of the mathematics contained in the work. How it came to be known is as famous as it is instructive to the student to always read the literature. In 1845, a fellow Cambridge student of mathematics and natural philosophy, William Thomson (later Lord Kelvin), was preparing to leave for France for an extended period of study in Paris when he read a notice in Robert Murphy's 1833 paper on inversion methods for definite integrals in the *Cambridge Philosophical Society* proceedings regarding Green's theorems. He received copies of the *Essay* from his tutor at Cambridge, William Hopkins (it is strange that despite Green's high ranking in the mathematical Tripos exam and his sojourn at the university he was almost forgotten by the time of this event). It changed his approach to physics and having found it, at last, Kelvin became the principal exponent of Green's ideas in England and Europe.

The striking feature of the potential function is that many problems become much clearer with this new mathematical representation of the force. In the *Principia*, Newton had shown that the attraction of a mass depends only on its geometry and not on its internal distribution. More important was his demonstration that for a homogeneous symmetric body, the gravitational attraction is due only to the mass located interior to the bounding surface. In spheres, for instance, this means the equilibrium of a body depends only on the mass located within any radius. This can be generalized for ellipsoids, a step first taken by Maclaurin. For instance, the shape of an incompressible body, a liquid, that is self-gravitating and in hydrostatic balance will conform to surface of equal potential. In this case, the boundary shape of the body, and its interior structure, are completely determined; the equilibrium surface is one of constant pressure (recall the implicit use of this principle by Archimedes). Thus, in the language now familiar from Green, the pressure is constant on equipotential.

There were many problems that could not be simply overcome. For instance, the question of whether the representation of the surface is unique was not clear. While it seemed that a distribution of sources would produce a unique force at any point in space it was not clear that the distribution could be found from the measurement of the forces themselves. The mathematical problem, that of boundary conditions and how they should be specified to solve the problem of the structure of the field, became a central area of research throughout the nineteenth and twentieth centuries. But using this concept of potentials did enlarge the range of questions that could be addressed in celestial mechanics. For example, the tidal forces felt by the Earth from the Sun and Moon had required complex formalism to resolve the forces in space and time. Now, with a single function representing the equipotentials the problem could be reduced to finding a single scalar function of position that, in an orbiting reference frame, could then depend on time. The oceans reside along an equipotential surface, precisely as do the charges in Green's idealized conductor, their surface being determined by hydrostatic balance. Therefore it is essentially a problem of geometry and analysis to compute the shape of the body and the forces that result on any mass within or on it. In this we see the beginnings of a much broader link between fields and force, the origin of inertia.

Why only gravity? There are other forces, so why was/is it so hard to make sense of this particular force? We return to the fundamental Newtonian insight: gravitation is a universal property of mass. The inertia is an attribute of a mass in motion or its resistance to acceleration, and therefore anything that changes the state of motion should have the same status. It is a force like all forces. The special place of gravitation is that it forms the background in which everything else occurs. But there are many such forces, not only gravity. In a rotating frame, the Coriolis acceleration depends on both the rotation frequency of the frame of reference and the particle velocity.

Now that the link between the gravitational force and the potential was clear the next step was to apply it to the structure of cosmic bodies. The problem had been examined before Green, by Newton in Book I of *Principia* where he demonstrated the necessarily oblate form of the Earth, and by Maclaurin. Green also looked at this problem. In all cases the studies used liquid masses, incompressible bodies in hydrostatic balance, so the form corresponded to the equipotential surfaces. The new wrinkle was a purely nineteenth-century conception. Having found that the bodies are in equilibrium under a specific range of conditions, for instance for certain rotation frequencies, the search began to understand if these bodies are stable. Since the time dependence of gravitational interaction could be either resonant or not for orbits, it was interesting to examine whether these different self- gravitating masses are stable. As the question of the origin of the solar system became more relevant, as the timescales became clearer (this was partly from geological reasoning and partly from thermodynamics), as it became clear that the system needed to have a mechanical explanation for its origins, mathematicians and physicists looked with increasing interest to the problem that became known as "cosmogony." The problem persisted. Nearly two centuries after Green, the study of the stability of self-gravitating bodies remained—and still is—unexhausted.

THE PRINCIPLE OF LEAST ACTION

The prehistory of the least action idea begins with Pierre Fermat in the 1660s who, in treating the passage of light through a medium, obtained Snel's law for refraction as a consequence of the principle of least time. At an interface between two optical media of different densities, if we take the length of a chord to be the sine of the angle, θ, the law of refraction states that the ratio of the chords on opposite sides of the interface, $\sin\theta$, is constant regardless of the incident angle and depends only on the change in medium across any interface, even a virtual one (for example, in the case of a continuously changing density). This can be stated as $\sin\theta' = n\sin\theta$ where n is called the "index of refraction." Now we take a medium of variable index of refraction and imagine that for any depth we can take some small distance over which n is constant. Then if we assume that to a depth in the medium the time to pass is given by the speed of light divided by the total length of the path, we can ask what is the minimum time to traverse a path or, conversely, what path will give the minimum time of passage or *least action* of the light? This is a variational problem, examining all possible paths irrespective of their curvature, to find the shortest total course for the light for a specified choice of the index of refraction. Fermat, in developing this variation principle, showed that the solution for constant index of refraction is a straight line and otherwise is a curve whose curvature depends on the depth dependence of n. You'll notice this requires knowing how to take infinitesimal steps to construct a continuous trajectory for a particle—including light—that required the calculus that was soon developed by Newton and Leibniz.

By the middle of the eighteenth century, physics had begun to spawn many interesting mathematical problems and methods, and this problem of least action was one of the most intriguing. Jean le Rond d'Alembert (1717–1784), following a proposal of Maupertius in the 1740s, made it the central principle for his derivation of mechanics. Extending Fermat's principle, he postulated that "among all paths a force can produce, the one followed by a particle under continuous action of a force

Figure 5.3: Joseph Louis Lagrange. Image copyright History of Science Collections, University of Oklahoma Libraries.

will be the one requiring least work." This is the one for which the total action, which is related to the *vis viva* of the body, has the minimum change in time. But how to define *action* was not so simple. It fell to Leonard Euler and Joseph Louis Lagrange (1736–1813) in the 1750s to develop this idea to its current form. It was later expanded by Lagrange and became the foundation of his analytical treatment of force problems in his *Mecànique Analytique*, first published in 1788.[4]

For Euler and Lagrange, this action principle became the foundation of mechanics and, therefore, a surrogate to force. Instead of thinking of action and reaction, the force is assumed to be derived from a continuous potential function whose spatial gradients determine its components. In two memoirs in 1760, and in subsequent editions of the *Mecànique Analytique*) Lagrange introduced a generalized variational principle based on action to obtain the equations of motion. We have seen with Leibniz that any particle with a momentum p will move in time with a *vis viva* whose sum over time is the action. Instead, if we start with a new definition for the action as a sum over infinitesimal changes in the kinetic energy during some interval of time. Lagrange assumed this quantity—what is now called the Lagrangian, the difference between the *vis viva* and the work—would be minimized among all possible choices of arbitrary displacements. Leibniz's definition of action, the product of the *vis viva* with the interval of time during which the motion takes place, is not the same as that adopted by Lagrange and Euler. The potential, that is the external force, doesn't appear and therefore this is only a conservation law for a free particle that is only valid for inertial motion. Instead, Lagrange and Euler explicitly included the effects of a background force. Since the work is the force multiplied by the displacement, the variational problem now becomes finding the way the kinetic energy changes such that the total variation in the work plus the kinetic energy vanishes. A rigidly rotating body also exhibits a "constrained" motion, its moment of inertia remains constant. For any mass, or system of masses, characterized by such constraints, Lagrange found they can be added to the variational problem because they remain constant. Take, for example, a system that maintains both constant energy and angular momentum. Then since each will have vanishing variation, the sum of the variations must also vanish. Although this formalism introduces nothing fundamentally new to mechanics, it clarified the link between force, in these cases weight, and *vis viva* (kinetic energy) through the applications of D'Almbert's principle and least action. But when introduced it provided substantial insight and, most important, heralded the application of small perturbations to generalized dynamical problems from celestial mechanics, thermodynamics, and eventually quantum mechanics. It is a change of viewpoint. To explain this further, consider that the problem of motion, not its stability, dominated the initial development of classical mechanics after the *Principia* for nearly a century. A vanishing net force is enough to insure inertial motion but it cannot provide the answer to the question "is that state stable?" In an increasingly broad range of investigations this became the focus. In particular, when in celestial mechanics persistent (secular) changes were obtained for the planetary orbits on relatively short (even by biblical standards, notably in the orbits of Jupiter and Saturn) it was imperative to determine if these were truly monotonic or periodic. The definitive answer for the Jupiter–Saturn system was provided by Laplace from a perturbative treatment: the changes are periodic but on very long timescales and the system is *secularly stable.*

For fluids the problem is more difficult precisely because they're *not* rigid. But remember Bernoulli's conception of an ideal fluid as a collection of streamlines. Each carries a mass density and has a volume as well as a mean velocity. They have, consequently, a momentum *density* and an angular momentum *density*. If we constrain them to not overlap, these streamlines can also form vortices and the motion can be treated analogously to the rigid rotation case. Instead of a body following a trajectory, the fluid is itself the trajectory since it is space-filling and the laws that govern the motion of single particles can be extended to a continuum. This was begun with Euler in the 1750s and extended through the following century as the concept of, at first an incompressible continuous medium and then a compressible one, developed. Here the problem of stability was also more difficult but the idea of using energy to derive forces instead of the other way around, to use the work performed by or on a system instead of asking questions about the external *forces*, gradually took center stage. For stability, a body must achieve its minimum energy and it must do so without changing its structure. For rotating fluid masses, for instance, this produced phenomena not encountered in single particle cases.

Hamilton and Action

The variational method was further exploited by Karl Gustav Jacob Jacobi (1804–1851) and William Rowan Hamilton (1805–1865). Hamilton's approach is especially revealing. He introduced what seems only a small modification to the Euler-Lagrange equations, taking as his *characteristic function* the sum of the *vis viva* and the potential, $H = T + U$, where, as for Lagrange, T is a function of both position and momentum but U, the potential, depends only on position. This function could be transformed into the Lagrangian by a simple rule, using the generalized conditions that the force—in the original sense of a change in momentum with time. Hamilton also wrote the velocity in a new way, as the change of this characteristic function with momentum. The surprising consequence of this purely formal change in definitions is a new framework for posing mechanical questions that is completely consistent with the least action principle. The equations of motion become symmetric: a time change of position depends on the momentum change of the function H and the rate of change of momentum with time depends on the spatial changes in H. This function, now called the Hamiltonian, is constant in time but varies in space; the motion is, therefore, constrained to maintain constant total H. This is a result of his choice of transformations between the coordinates but there was something more here that Lagrange had not realized. These methods were remarkably fruitful, so much so that in the twentieth century they were the basis for the development of quantum mechanics and field theory, as we will see. They also served for celestial mechanics in an important way. They permit a transformation between measured properties, such as angular motion, to more general properties of orbits, such as angular momentum. At the time Hamilton was concerned with optics, specifically how to describe wave propagation mechanically. For a simple wave, the frequency and phase of oscillation provide a complete description of the force. For light the propagation speed is already known and he realized that a plane wave is defined by a surface of constant phase. He identified this phase with the action, he called this the *eikonal*, and by assuming that the trajectory followed by the wavefront is that which minimizes its change in phase, recovered the least action principle.

The utility of the Lagrangian approach, and the later generalization by Jacobi and Hamilton, is its transparency when changing from one coordinate system to another, especially when there are constraints. This had already been introduced in celestial mechanics by Poisson. He realized that instead of working with the dynamical quantities, position and velocity components, it is more natural to represent orbits in terms of measurable geometric quantities such as the eccentricity, orbital period, and eccentric anomaly. He showed how to transform the equations of motion using these parameters and his mathematical device was essentially the same as Hamilton and Jacobi found for passing between different reference frames and conserved quantities. There are a few examples that will show how this new approach could provide answers to mechanical problems that otherwise were quite cumbersome. One of the most beautiful comes from examining a device that was used to illustrate ideas of equilibrium of forces, introduced as a schematic machine by John Atwood in 1784. It is the simplest machine, the descendent of Jordanus' idea of *positional weight* except in this case the masses are connected by a cord draped over a pulley (whose mass we will neglect) hanging aligned and vertically. If the two are equal, they can be placed in any relation with respect to each other and if one is set into motion, the other moves in the opposite direction with precisely the same speed. If, on the other hand, the masses are unequal, they cannot be placed in any relation that will allow them to remain stationary.

INTERMEZZO: THE STORY SO FAR

Let's now sum up the developments of the force concept at it stood by the middle of the nineteenth century.

For Newton himself, the pivotal moment came with the realization that, according to his conception of force, any deviation from inertial motion must be in the same direction as the force provoking it and of an equal magnitude. Thus, for any mass, any change in the path is the result of an external source producing its acceleration. If a mass falls, it proceeds toward the center of the force. This means, here, toward the center of the Earth. because, then, the Moon orbits the Earth and there is nothing else changing its motion, it must be able to fall precisely as a rock at the surface. Being farther from the center, the influence of the Earth—the force per unit mass and thus the acceleration—will be smaller but the same terrestrial rock would behave just like the Moon at the same distance from the center. That the motion in angle is different is easily dealt with. A rock thrown horizontally would, asymptotically, simply never hit the surface as its "inertia" increases, it would continue to fall without ever hitting anything. If the Moon, in this same sense, were to stop in its motion, it would fall and take the same amount of time to reach the center as it does to orbit because the freefall time is the same as the orbital period. Again, there is no angular forcing, no torque.

Now for the next step, we say that the only agent producing this is the mass contained within the orbiting path. Any outside has a perturbative, deviating effect unless it is identically distributed as the central body. In the symmetric case, there is no additional influence, apart from some possible viscosity (if this is a fluid, for instance), only its mass within the circle enclosing the orbit would directly affect the orbiting body and this identically as if all of the interior mass were concentrated at a point in the center. The power of this realization was immediate: all bodies orbit the Sun following the same law because it is the dominant force. For the Earth–Moon system, the Earth has the dominant influence on the Moon, as on the Rock, but the two orbit the Sun because the differential force of the Sun across the lunar orbit isn't sufficient to make the Moon also orbit only the Sun. This produces, however, a phase dependent force, one that changes throughout the month (lunar orbit) because it changes phase through the year. The same for the tides, the differential force across the Earth is large enough to produce a relative displacement that depends on phase for the two sides of the Earth.

The dynamical theory of tides then takes the inertial reaction of the masses into account, allowing for internal forces to change the motions, but the static theory was completely developed subject only to the distortion produced on the body in equilibrium. Allowing for a different reaction time for the fluid and solid parts of the body produces phase shift between the force maximum at any point and the reaction. This was the step taken by Laplace.

Newtonian gravitational theory was, however, only concerned with the forces and therefore the change in viewpoint with the potential function was to add a means for deriving the forces from the mass distribution. Independent of the symmetry of the body, the forces were locally produced by changes in the potential along the direction normal to any surface. The geometric form of the body thus produced a concentration of the force, or a reduction, depending on the gradient of the field. Again a geometric device replaced a physical explanation, it doesn't matter what the field is, any field will behave identically if the force is given by a field. The symmetry produces certain constant quantities, for instance in a spherically symmetric or axially symmetric system the angular motion remains constant because there is no gradient in the potential along the direction of motion. But if there is any change in the form of the body, a broken symmetry, that alters the angular as well as radial motion and the orbit will no longer close. The motion of the body, whatever the source of the field, will alter.

The last step was to take this static picture, where the masses or charges were not changing their positions in time, and allow for the source itself to vary. In particular, if the geometry of the source changes in time, this produces a change in the reaction in time. For a gravitational field this is a gravitational wave. For an electric field, a charge distribution, this is an electromagnetic wave. Mass currents are different than charge currents but not so different at this basic level. A wave is a time dependent acceleration at a point in space. And the inertial response of a body depends on its coupling to the time variation, if there is an angular component to the wave this produces an angular response, if it is strictly radial this produces a simple harmonic oscillator in radius and the body doesn't displace relative to the central source. Thus, for Newtonian gravity, there is only instantaneous change, the effect of introducing a spacetime is that all changes require a finite period and thus resemble electromagnetism. There are waves, polarized and accelerating, for gravitational fields exactly as for electromagnetic and because currents and time dependent displacements of charge are of similar character, electromagnetic waves require both changes in the magnetic and electric fields.

NOTES

1. For the rest of this book, we will describe this as a "scalar" quantity, something that has only magnitude, for instance a length or speed. In contrast, a "vector" quantity has both direction and magnitude, such as a force or a velocity. The terminology was introduced by Hamilton but the ideas are as old as the composition of motions and the intension of forms.

2. To say this in more precise language, we take the field strength to be the gradient of a function, Ψ that depends only on space so that $\mathbf{E} = \text{grad}\,\Psi$. Then the divergence of this quantity, summed over a completely enclosing volume, is

$$\int \text{div}\,\mathbf{E}dV = \int \mathbf{E}\cdot\hat{\mathbf{n}}dA.$$

Here V is the volume for which A is the bounding area and $\hat{\mathbf{n}}$ is the direction of the surface normal. The divergence is scalar in the sense that it is a quantity, a rate of change in all directions of something, while the gradient is a vector since it is the rate of change along each direction. By dimensional analysis the field strength varies as the inverse of the area times the charge within that distance so the volume integral of the density is the total enclosed charge. This results in the Laplace equation for a point charge and the Poisson equation for a distribution.

3. Let me advertise the discussion of relativity theory in Chapter 8. You'll see why this way of describing a field was so important for Einstein during the first stages of his groping for a way to express a generalized theory of gravity using curvature, hence a geometric framework, is natural for the description of the field. His use of the word potential, or field, for the metric—the way to measure the structure of the space—is not a casual choice: if you think of the potential as a surface, and the motion is therefore expressed in the extended framework of a four dimensional world instead of only three, if the time and space are not really independent of each other, then the acceleration can be understood in terms of gradients of those surfaces and in terms of their curvature.

4. This work is famous for its preface, which declares "One will not find figures in this work. The methods that I expound require neither constructions, nor geometrical or mechanical arguments, but only algebraic operations, subject to a regular and uniform course." With this treatise, Lagrange was continuing the mathematization program begun

by Galileo but with a radical modification. While previous arguments had used geometric formalism and presentation, the *Mecànique Analytique* explicitly adopted purely formal language. This approach was to take root on the Continent rather quickly, it took several decades of propagandizing in the first decades of the nineteenth century by Charles Babbage, John Herschel, and George Chrystal in England to affect the same reform.

6

THERMODYNAMICS AND STATISTICAL MECHANICS

Birds of a feather flock together

—English proverb (c. 1545)

Science and society converged at the end of the eighteenth century with the beginning of the "'Industrial Revolution," the development of factory-based mass production of identical goods at high rate and of almost unlimited variety. Mechanical invention provided the impetus for this momentous shift from small scale to industrial mass production and with it came a new concern: efficiency. Central to this change was the steam engine, in its manifold forms.

The start was innocuous enough. Coal recovery was frequently hampered by flooding, a problem that could be alleviated simply when the mines were shallow by ordinary air pumps acting against atmospheric pressure. But there is a limit to such work, a column of water cannot be raised in single steps by more than about ten meters. Galileo, Torricelli, and Pascal had known this two centuries earlier. Georg Agricola, at the end of the sixteenth century, had discussed its practical consequences. For deep pits, many pumping stages were required to elevate the water. To make the process more efficient, some force had to actively drive the fluid to greater elevations. This required large pumps and the most efficient driver was a battery of steam engines.

The second was the problem of factory layout and operation. One skilled artisan working through a complete operation was too slow for the industrialists. Instead, the separation of individual tasks required specific machinery for each and the ability to pass work partially completed to another hand for finishing. Water power was enough for the older collective producers. But for the expansion of scale needed to satisfy ever-increasing consumer demands, a more machine-centered mode of production was needed. Water and wind power, both of which sufficed for smaller scale manufacture, face severe limitations when scaled up to the level required for the new factories. There is no power to drive a wheel if there is no flow; a dry season could be a financial disaster. A wet season, on the other hand, could prove equally devastating since often there were no reliable mechanisms for controlling the rate of flow. This hydro-power also required mills to be placed at specific locations. Further, the power could not be easily distributed within a large factory complex.

The industrialization of Europe was advancing apace and increasingly required technological improvements to improve the efficiency and reliability of its machinery. In this context, the growth of thermal physics was a natural, almost inevitable outgrowth of a social transformation. The new questions raised by the needs of the industrialists provided fruitful and challenging problems for the mathematicians and scientists of the era. In particular, in the context of the Newtonian program and philosophy of the regulation of natural processes by natural laws, there was an open territory for exploring new theoretical and applied ideas. Human labor was difficult to quantify and even harder to control. The machine provided important benefits to the new "captains of industry" and the capitalists who financed them. One was the ability to precisely measure the *efficiency* of the process, the amount of fuel required (in whatever metrical units one would want, for example cords of wood or tons of coal) and some measure of the productivity or output work, for instance the amount of cotton separated and combed or the area of woven cloth. As the devices, mainly steam engines, were put into progressively wider service, the measures of the work achieved varied as did its physical manifestations. Overall, however, it was becoming clear that Nature itself operates with some sort of balance sheet, totaling the costs and production to lead to profit.[1]

There are many examples, I'll choose just a few. The expansion of steel production was spurred by large projects such as the construction of metal ships, bridges, and buildings. While much of this could be carried out using purely practical techniques, such as those used by the medieval master builders in the construction of the Gothic cathedrals, national-scale projects such as the creation and expansion of railroads during the first half of the nineteenth century throughout Europe and North America and the hydrological studies required during the era of canal construction and the increasing needs for water supply to expanding cities, all produced auxiliary theoretical problems of generalization and extension that dominated much of physics during the nineteenth century. It's important to keep this in mind. The expansion of physics, chemistry, and mathematics during the century was frequently spurred by complex, practical problems that required general laws rather than "rule of thumb" approaches.

Above all, the invention of the steam engine was a fundamental historical turning point in technology and physics. While during the previous centuries ballistics had provided a (unfortunately frequent) context for posing general questions in dynamics, the newly founded centers of industry, whether Manchester in England or Clermont-Ferand in France or the Saar in Germany, also required a growing presence of technically trained engineers and designers. This in turn required an expanded university presence for science, especially physics, and it was during this period that many new universities were formed, and the curricula changed to emphasize the teaching of physics at all levels of schooling.

The steam engine provided the framework in which to explore what was quickly recognized as an obvious connection between heat and force. One way to think of how the picture changed is to recall a level or a block and tackle. These simple machines converting in effect, one form of force into another and as such had been classified by Hero of Alexandria and Vitruvius soon after Archimedes. But a steam engine is very different. It transforms heat into work, taking energy—not simply force—from the state changes of a hot vapor and converting that to mechanical action with some efficiency that can be quantified. Although the essential force laws for vapors and liquids had been explored in the seventeenth century—for instance, Boyle's law for the change of pressure in a gas on compression—the thermal contribution had been neglected. Even the atomists, for instance Daniel Bernoulli, had pictured a state of agitation of microscopic particles that didn't involve a change in their momentum distribution; the particles merely impact the walls more frequently when the volume is decreased but that's purely a density effect. Instead, to consider changes when a quantity of heat is added, was in the program of the nineteenth century and was spurred by the development by Newcomen and Watt of engines capable of driving pumps, lifting heavy masses, and running whole factories.

It was a development in part directed by an economic need. A steam engine consumes a quantity of fuel. This costs money. It produces a measurable output, work, that represents a certain income. Far more than human or animal labor, where the fuel source is far less immediately understood, anyone can understand how much petroleum or wood must be burned to produce a specific output of the machine. Thus was formed the research program of thermodynamics. Heat is an internal property of an object, something that can be quantified with a thermometer (any device will do). But work is the measure of its interaction with an environment. A piston expands to push a level to drive a wheel. The momentum of the wheel is the result, then, of a force originating at the piston. The fuel, steam, changes state—it cools—thus releasing a certain amount of received heat, supplied by a fire, to a motion. Yet state changes were also involved. A column of liquid, mercury or water, expands when heated. That is, simply imagining a vertical column standing on the surface of a table, the change in the height is a work done by changing something internal to the fluid. The implication that something has been transferred to the fluid from the surroundings, perhaps a fire, that without any distant action changes the state of the material. But how do you define the efficiency of any device for converting energy—here in the form of heat—into work, and how do you describe physically what happens.

The dream of the industrialists was to get something for nothing, to get more work out of their machines than the fuel supplied. Motion once started might continue forever, that's what inertia means. But real devices have dissipation so, given these losses, is perpetual motion—or better, perpetual action—possible? As early as the late sixteenth century Simon Stevins had considered the perpetual motion of a continuous chain hanging under its own weight on an inclined plane. He asked whether this "machine" could continue to move or, for that matter, would start moving, abstracting the essential dynamics from this rather cumbersome driving. His idea was to contrast motion on an inclined plane with freefall, hence the choice of a wedge rather than something more realistic. The answer was no, and the same reasoning applied in Galileo's discussion of inertia. These were relatively simple to address since the equivalence of actions of weight independent of the constraints of the geometry. But when it came to mechanisms driven by heat the answer wasn't so clear.

CHEMICAL FORCES

To understand how to transform matter into work required knowing what forces govern its states and reactions. For Newton, chemical combinations and transformations served as microscopic analogies to the forces and their transmission on large scale. But his speculations, such as those appended to the *Opticks* between editions of the *Principia*, were far from quantitative. In the last third of the eighteenth century and the first decades of the nineteenth, chemistry became a systematic science in the hands of Antoine Lavoisier and John Dalton. Chemical systematics were discovered, along with the existence of substances that despite distillation could not be reduced, the elements. Lavoisier showed that oxygen is the active agent releasing heat in chemical reactions in air and Dalton discovered the combinational rules for weights of the elements. With the demise of a special fluid to explain combustion and the discovery of integer relations among elementary volumes and weights of the elements, chemists turned again to the atomic conception of material structure.

Recall that although the atomic picture was a ancient inheritance, it came along with an antique mechanism to explain the structure of matter. All forces were contact and specific to the nature of the particles: hooks and eyes, sticky surfaces, and analogies familiar from the artisan's workshop. As pure analogies these worked perfectly to "preserve the appearances" but they were useless for predicting new phenomena. Even more extreme was the hard elastic sphere used by Boyle and Bernoulli to explain the gas laws. Again, the hypothesis explains the observations but the model leads does not suggest new phenomena. Transformations of the sort so familiar within alchemy and later in experimental chemistry required a new approach and this was provided by Boscovich in his *Theory of Natural Philosophy* (1758). In the spirit of the mechanical philosophy, he introduced a sort of field theory into the atomic world, picturing the interactions among atoms as both attractive and repulsive according to the specific *elements* involved. This was not just a hypothesis. By the middle of the eighteenth century it was clear from generations of exploration that there are a number of substances that, despite being subjected to continued treatment with heat and distillation, cannot be further purified. These were called *elements* and the study of their properties founded not only modern chemistry but also renewed the atomic view and brought the study of matter into the proper domain of physics.[2]

In one of the most fruitful scientific collaborations of the eighteenth century, Laplace and Lavoisier studied the properties and chemical reactions of solutions and introduced some of the basic concepts that later dominated chemical thermodynamics. They imagined immersing a beaker in which two solutions had been mixed within a huge block of ice, or better a thermos-type bottle in which instead of a vacuum the space between the two cylinders is filled with packed ice. When the reaction occurs heat is released. This is transmitted to the ice, which melts, and the water is drained. By weighing the quantity of water, knowing the specific heat, the amount of energy released can be determined. This is a system that changes energetically but not mechanically, in effect this is a system that does no net work but changes its internal state, its heat content. Importantly, at this state change isn't reversible. Another example is to think of a rock. You lift it and do work on it against a potential field, the background gravitation. When released it will have gained no net energy, no net work can be done on the system. But when it hits the ground some of the potential energy is un-recovered, it goes into heating the body, perhaps breaking it. You may be able to use the rock later, if it's hot enough (try to imagine that) to heat something else but it will not spontaneously cool and jump into the air. In both cases, we're really confronting the earliest concepts of potential versus kinetic energy. But the two, and many others, are also illustrating the difficulties of maintaining such motion under less than ideal conditions. If there is any friction, between rubbing solids and surfaces or internal fluid viscosity, the motion will clearly cease.

MACHINES AND THERMAL CYCLES

The idea of using heat and steam power as a driver is as old as the Greeks. Hero of Alexandria described a wonderful device for opening temple doors when a sacrificial fire was lit on an alter. But the development of a machine driven by burning coal or wood to heat a boiler and drive the expansion and contraction of a piston was the product of the eighteenth century and changed the world. It began in those same mine shafts I'd mentioned a moment ago. As the need for coal grew, miners descended ever deeper and the problem exacerbated. The first practical steam powered engine, invented by Thomas Newcomen in 1712, was put into service as a pump. The difficulty with the original design was a tendency of the machine to go berserk, to overdrive and rupture. The valved condenser and dynamical control were invented by James Watt (1736–1819) in 1769. For the condenser, the expansion and evacuation were moved to a separate container. For the control, he attached a spinning ball, suspended on flexible supports, to a shaft that was, in turn, connected to the driving shaft of the main wheel of the engine. In this very neat mechanical invention, if the wheel turned too quickly the balls rose and extended, increasing the moment of inertia and slowing the device much as a skater does in extending his arms. If the rate fell, the balls dropped and contracted below their balance point and the centrifugal force reduced. This feedback mechanism, called a "gouvernor" for obvious reasons, made steam power practical. Thus the steam engine became the paradigmatic device to illustrate the later discussions of dynamical thermal processes. While the concept of heat was still ill-formed, the load of wood (and later coal) needed to fuel the engine made it clear there was no *perpetual work*. But the discussions of the mechanisms were still confused. While force was obviously produced because the pump was clearly doing work in lifting a weight (in this case of the liquid), and this came at the expense of the supplied heat, it was far less clear that there was actually a *conservation principle* here analogous to that of momentum.

HEAT AND WORK

A force can be produced by, or can produce, changes in the "state of a body" that we call heat. How this internal property is changed into an external acceleration and/or work, is something that perplexed the physicists of the nineteenth century and still produces confusion. You see, the motion of a particle being acted on by an external force is a comparatively easy thing to follow relative to the change in the condition of the body or system being thus affected. Work, the production of a mechanical effect, is something that can be quantified but the relative balance of energy among different components, or types, is much more difficult to ascertain. For instance, how much heat is added to a body can be measured only if you can control the state of the system at a particular time.

Let's stay with the steam engine for a moment (although you could imagine this in any generator, the treatment of a nuclear reactor is virtually the same except for the fuel). The operation is something like this, in general. Place a cylinder filled with water in contact with the heat source, a coal fire for instance. Absorbing or releasing heat is a different from doing work. When in contact with a heat source, the gas temperature rises in the cylinder but there is no work. Any addition of heat produces both a change in the internal energy. Now take the body away from the heat and allow it to expand. This will produce the work since the cylinder, if properly constructed, drive a lever and through that anything else in the outside world. The machine does work. But there is a price. The internal energy decreases, the cylinder cools. Now, once we reach a specific temperature, we stop the mechanical process and allow the rest of the heat to be released, perhaps even involving flushing condensed water out of the cylinder. Then we compress the gas further in isolation, and prepare it to be again placed in contact with the heat source.

A theoretical framework that defined the efficiency of a thermal engine was first proposed by Sadi Carnot (1796–1832) in his 1824 memoir "On the Motive Power of Fire." Carnot's approach was to consider a ideal machine, an abstract device without considering its construction or composition, operating between two fixed temperatures. These determined the quantity of heat transferred to it from a reservoir. In this schematicized picture the bath in which the device is immersed, or its sources and sinks, are irrelevant to the operations. Its sole function is to convert a difference in temperature, produced by the absorption of heat, DQ, into mechanical work, W. As a measure of the *efficiency*, he proposed the expression W/DQ, the work done per unit heat absorbed. He imagined a reversible process, or a sequence of them, that not only operates between fixed states but occurs so slowly that the device can remain in equilibrium at each point in its changes. By this Carnot meant again to draw the comparison with a steam engine but one that moves so slowly that at any moment the entirety of the device is at the same temperature and there are no net accelerations in the system. In this case, during the interval when things are changing, there is no exchange with the sources or sinks. This occurs only at the ends of the process. For a cylinder of a steam engine, this is a cyclic change. The system absorbs heat, does work, and them prepares itself by a mechanism to again receive some quantity of heat. This is now called the *Carnot cycle*. The cycle can be subdivided when operating between any set of temperatures such that the extrema represent the global change in the system while the subcycles can be considered as taking the system between a set of intermediate states each of which is only virtual.

Carnot used indicator diagrams to illustrate the process, a graphical device introduced by Watt to describe the cyclic operation of the steam engine in which the state of the system is described by two easily measurable quantities (such as temperature, pressure, and volume). Various combinations were employed, depending on what properties were essential for describing the machine's operation. One could choose these *intensive* variables (those not subdivided within the device itself), the temperature and pressure were the principal physical quantities, and the volume of the device since the work of a steam engine is done by successive expansions and contractions. There is a stage when condensation occurs because of the expansion, the part of the cycle that ultimately limited its work, because of the collapse of the steam to liquid when some critical temperature is reached on expansion. But that was expelled from the system, a waste product, which is why the machine cannot be completely efficient. The beauty of this abstraction is that you can ignore the structure of the device. In fact, any thermo-mechanical system, no matter how complicated, can be treated as one composed of parts in contact. This was later generalized to include the idea of a heat bath, a complete environment in which a body is immersed, and made more precise the idea of equilibrium.

To then extend this approach we note that it must be true for any system if the path of the transformation doesn't matter so we are free to choose the description of the material. Benoit-Pierre-Emile Clapyron (1799–1864) and Rudolph Clausius (1822–1888) chose to quantify the effects using an ideal gas, one for which the pressure depends only on the temperature and volume. Then because the internal energy of the whole cylinder depends on the temperature alone, the transformation of heat for a reversible process is a smooth one that has *the change produced by the addition of a quantity of heat to a thermodynamic system produces work and changes the internal energy*. This is the first law of thermodynamics, the relation that defines the *mechanical* content of thermal physics. Clausius, in particular, distinguished two energies of any mechanical system. One is the center of mass action, the bulk motion generated by the potential or force. Whether this is a pressure acting on the device or a force on its center of mass doesn't matter. But to expand this definition meant accounting for both external and *internal* changes and, within a body, the energy is quickly (it was assumed) redistributed in a wide spectrum of motions, none of which can be individually resolved. None of these changes produce any net effect on the body as a whole, it continues to move as it had although its shape may change because of internal rearrangements. Thus the center of mass acceleration represents the reaction to a potential, while the change in the internal energy—what Clausius described with his title—is "a kind of motion we call heat." While it is possible to use this waste product of the average mechanical process to do other things later, it isn't available to perform the original function of the device.

HEAT AS A FLUID

Steam engines and other thermally driven machines are conversion devices taking something from the fire and converting it into work. There was an analogy here with Franklin's conception of a single fluid for electricity (although this had been superceded by the end of the eighteenth century, as we'll discuss in the next chapter), a thermal fluid that is transferred from something having an excess of the fluid to something else that's deficient. To make the analogy clearer, we imagine a quantity of something we'll call heat, *caloric*, that can be transferred precisely like a fluid: it is assumed to be infinitely divisible (not particular) and is measured by something we will call temperature. To say something is hot, instrumentally, is to take a reading with a device—a thermometer—and notice that the reading is larger when we think the body is hot, as Carnot says, which *is* how there is much heat in a substance. While we really don't yet know what this means, we can imagine changing the amount by DQ. Then there are two effects and this is the *first law of thermodynamics*, that the quantity of heat added to a system in contact with a source produces a change in the internal energy plus some work done by the system on its surroundings (or anything else to which it's mechanically coupled). This is a *law*, on the same footing as Newton's first law defining inertia. As long as a body is in contact with a source of heat, work can be driven but the body will also get hotter.

Now in keeping with out definition of work as the action of a force during a displacement by that component lying along the direction of motion, we can also imagine an extended body changing. Here we don't ask detailed questions about the internal structure, simply that there is a way of describing how the body as a whole will react to the addition of this heat. The body has a volume (or a length, this argument applies equally to the change in the length of a rod, for instance, if heat is added) and, therefore, a surface, and the pressure is defined by the force distributed over an area. If we sum over all the areas and take the increase in the volume to depend only on a lengthening of the column (imaging a cylinder, only one end changes while the walls remain rigid). Then the work, which is the force times the displacement, becomes $W = P\Delta V$ for some infinitesimal change in the volume, ΔV if the pressure is constant during a period of very small change in the volume. We'll return to this point in a moment. Remember that Carnot imagined that the changes are taken slowly and in infinitesimal steps so there is sufficient time in each step to reestablish mechanical equilibrium. This required that the input of heat produces a change in the internal energy E and the work W ($DQ = \Delta E + W$); E is that part of the heat that doesn't produce work or a change of state of the body. The way we change the system matters, and this is where thermodynamic reasoning is so different from ordinary dynamics. If we hold the volume constant, or we hold the pressure constant, we get different results for the change in the internal energy. If we consider the thermal analog of mass to be the heat content divided by the temperature, $\Delta S = DQ/T$, then we can imagine the change in the heat being a simple scaling number. If the heat capacity of a body is constant then DQ/T is independent of temperature and therefore should not vary if the changes occur slowly enough that no net work is done, an *adiabatic* change. Now let's go back to Carnot's hypothetical machine that simply operates, somehow, between two temperatures, and ask what amount of work we can get out of the device efficiently. Let the quantity of heat initially absorbed be Q_1 at a temperature T_1. Then let the system expand without loss of heat. Then imagine the system to be cooled, or to perform work—it doesn't matter for our purposes—but the amount of heat lost should be the same as the amount gained except that it occurs at a lower temperature T_0. Then let the system contract, somehow, until it is brought again into contact with the heat source. The total work done will be proportional to $(T_1 - T_0)\Delta S$, the amount of heat absorbed will be $DQ = T_1\Delta S$ so the efficiency will be $\eta = (T_1 - T_0)/T_1 < 1$. In finding this result, that it is impossible to create a perfect machine, Carnot demonstrated conclusively that perpetual motion cannot exist even for a purely thermally driven system.

MECHANICAL EQUIVALENT OF HEAT AND HEAT CAPACITY

The specific heat is defined as the coefficient for the change in the internal energy—that is, the temperature—when a quantity of heat is added but what "the amount of heat" means was ill-defined without some relation to work and/or forces. This problem was solved by determining the *mechanical equivalent of heat*. The concept originated with Count Rumford (Benjamin Thompson, 1753–1814), as often happened in the early history of mechanics, through a practical observation. You know that rubbing your hands together produces, through friction, a feeling of warmth. Repeated bendings of spoons produce substantial heat and even eventually leads to it breaking. Stressing a rubber band between your lips will produce a sensible temperature difference. Rumford noted that an enormous amount of heat was generated by friction in the boring of cannons, and communicated his observations to the Royal Society in a series of influential memoirs. These common experiences make you suspect there is some connection between work and heat but aren't quantitative. That was accomplished experimentally in an elegant way by James Prescott Joule (1818–1889). He inserted a paddle-wheel in a water-filled sealed cylinder and connected the mechanism to an external hanging weight. He used distilled water, which was the standard substance in the nineteenth century because of its stability at high temperature and the ease with which it is purified. The centigrade temperature scale was calibrate by Celsius using its two well defined phase transitions: when pure water freezes and when it boils. Joule let the weight descend and measured the temperature rise of the water. The work done in dropping the weight depends only on the distance through which it is moved, not the path or the acceleration (which is always gravity). Once you have this quantity, the specific heat can be defined as the rate of change of heat content with temperature, $C = DQ/\Delta T$. The precise value of this depends on the substance. But how to relate it to the properties of matter was the beginning of a separate study, the statistical description of motion on the microscale.

EXPLAINING THERMODYNAMICS: STATISTICAL MECHANICS

Thermodynamics was about dealing with heat, energy, and work. For all the thermodynamicists knew, these could be fluids that are exchanged between elements of any complex system. The chemists were, instead, more concerned with combinational regularities and to them the world was increasingly obviously atomic. Mineralogy also revealed the connection between forces and structure at the microscale. While atoms in motion are difficult to imagine, and even harder to deal with theoretically because of the statistical element of their interactions, crystal structure is a clean example of how the forces between atoms can be understood. Different elements, carbon for example, structure themselves in a limited set of structures, the lattices that echo the Platonist vision of the "ideal, geometric" underpinnings of physical laws. Crystals display geometric regularities that fell into a small number of point group symmetries that also seemed to require some sort of microscale discreteness. Finally there were phase changes in matter, transitions between solids, liquids, and gases, that required or released heat accompanied by often dramatic changes in volume without changes in weight that were consistent with an atomic model. These growing conflicting views about the nature of heat reflected a more fundamental shift in dealing with matter itself. The interaction of atoms among themselves was, in Gibbs' view, the reason for phase changes. A *chemical potential* measures this and even in the absence of motion and at both constant pressure and temperature, reactions change the entropy of the system by altering the number of particles. An increasing collection of observations and experiments were, by the last third of the nineteenth century, pointing to both an atomic view of microscale interactions and a discretize picture of the internal structure of those particles. These realizations arrived from many different directions, some from the laboratory, some from natural phenomena, some from theory.

The Virial Theorem and Phase Space

Thus, during the second half of the nineteenth century, heat ceased to be a fluid, or a field, or something substantial. It became, instead, *a dynamical property of matter*. It was something within things that connected the structure of material with energy. Since matter and motion, for the previous 150 years, had become progressively more unified, the stage was set to add thermodynamics to this. But it was clear that something also had to change regarding the usual mechanical ideas for interactions of masses. Thermal properties are *not* universal in the way gravitational interactions are. Different compositions produce different behaviors. Even for the same material, depending on its state the thermal properties are different. The gas laws and chemical systematics discovered Lavoisier, Dalton, and their contemporaries between about 1780 and 1820 seemed to demand an *atomic* view of matter and with it came the question of whether the dynamics of atoms are different from those of macroscopic bodies. How is this microworld structured? This became a central question of physical theory in the middle of the nineteenth century. The laws governing the motion of atoms were constructed analogously to those governing gravitationally interacting masses. Although gravitation plays no role, and this was quickly realized, the idea that interactions can take place at a distance through the action of a field dominated the modeling effort.

The modern concept of *statistical* mechanics was founded in the second half of the nineteenth century, mainly by James Clerk Maxwell (1831–1879), Ludwig Boltzmann (1844–1906), and J. Willard Gibbs (1839–1903). The *virial theorem* of Clausius was an important step in this program. Although it was implicit in the dynamics of macroscopic, gravitationally dominated systems of masses, Clausius realized that it could be generalized and applied to any system in equilibrium. He realized that since the combined potential and *vis viva*—kinetic—energies taken together are the *total* energy of a body, he extended this concept to a system of bodies mutually interacting through *any* potential field. The resulting statement is that the sum of all kinetic energies K of the particles was half of the total energy, or that the binding energy of the system (in equilibrium) is half of the total energy. This is for a linear system, one where the potentials Ω depend only on the relative positions of the constituent particles, written as $2K + \Omega = 0$ if the system changes so slowly that changes in its moment of inertia can be neglected. This is a global constraint on the system, the trajectories of the individual constituents don't matter. Joseph Liouville (1809–1882) had derived a related constraint that for an adiabatic change, the distribution of particles would shift around in space and momentum and exchange energy but maintain constant energy.

This is why the dynamical picture developed by Lagrange and Hamilton was so important. Every particle at any instant in time has a defined position and momentum. Each of these has three spatial components, since they're vectors. Taken together for N masses they form a 6N dimensional space, called a *phase space*. You'll recall this from the discussion of celestial mechanics, orbital motions and perturbations are described in terms of orbital elements that are functions of the energy and angular momentum and the equations of motion are the rate of change in time of the positions and momenta with changes in momentum and position, respectively, of the Hamiltonian. For a gas, we eschew the possibility of tracing the motion of any single particle and instead assert that we only measure *mean* quantities of a statistically distributed set of positions and momenta. We imagine a collection of masses each occupying at any instant a point in this space but, because of collisions and distant encounters, the coordinates randomize and say they behave probabilistically. What we see is an *average* over some statistical distribution, we'll denote as $f(\mathbf{x}, \mathbf{p}, t)$, which is the probability of finding any single mass in the ensemble at a specific point in phase space. For a uniform distribution in which no spatial coordinate are correlated, we can write $f(\mathbf{x}, \mathbf{p}) = f(\mathbf{x}, p_1)f(\mathbf{x}, p_2)f(\mathbf{x}, p_3) = n(\mathbf{x})f(\mathbf{p})$, where n is the number density.[3] Then if the momentum is identically distributed in all directions and the total kinetic energy is the sum of the partial energies (that is, $= (p_1^2 + p_2^2 + p_3^2)/2m$ for identical particles of mass m) there is only one function that satisfies the requirements. This function was found in 1859 by Maxwell in his Adams prize essay on the stability of Saturn's rings and made more precise in his papers of 1860 on the dynamical theory of gases: it is the Gaussian or "normal" distribution that was already known from statistics, $f(p) \sim \exp(-p^2/(2mkT)$ in an interval of momentum $dp_1dp_2dp_3 = 4\pi p^2 dp$ (since each interval of momentum is also identical) where k is a constant, now named after Boltzmann. Thus, if the energy is constant in the sense that the collisions do not produce losses from the system, we can change the average from the relative motions to the individual energies and write the distribution in terms of the kinetic energy $f(E)$. Each particle has a kinetic energy of $kT/2$ for each degree of freedom.[4] We can say nothing about the individual trajectories but we *can* make statements about the ensemble and these are the thermodynamic quantities. The biggest change in viewpoint is the reversibility of the action and reaction and the difficulty of assigning an order to the effects. Instead of solving the equations of motion for a single body, we imagine a collection that spans the full range of possible positions and momenta but nothing more. The equations of motion for macroscopic bodies can be explained by this picture. The atomic world of this classical physics requires more a shift in focus. Our interest is now on the interactions and not the trajectories, on how the forces that govern the microworld average to produce the observable behavior.

Brownian Motion

A glimpse of the microscale was provided by a simple observation by Robert Brown (1773–1858) when he described the motion of pollen grains in a liquid viewed with a microscope. The grains moved unceasingly but irregularly as if alive. This was not entirely unexpected, impulses had been used since Newton to describe how force is transmitted to the mover, but now it was different. Instead of the deterministic motion generated by a central field, this motion is random. This was Albert Einstein's (1879–1955) first fundamental contribution to physical science. In a series of papers in 1905, he solved the problem of how an atomic scale process can produce visible displacements. He assumed the cause of the motion is a continuous, enormous number of infinitesimally small impacts by the atoms of the medium in which the larger particles were immersed. Each impact can be in any direction and over an unbounded range of strengths, but the most probable are vanishingly small. To account for this, he assumed a Maxwellian distribution for the velocities and showed that the motion describes a diffusion equation by treating the motions as fluctuations of a continuous property, the change in the momentum of the particle. Each impact produces a deviation. In a random distribution of changes in velocity, the most probable value for Δv in any direction is zero. But there is a dispersion of the fluctuations and this never vanishes. For an ideal gas, as we have seen, it depends on the temperature. This prediction led to series of delicate and beautiful measurements by Jean Perrin 1870–1942) in 1909 that traced the trajectories of particles and quantitatively verified Einstein's law. Paul Langevin (1872–1946) generalized this by writing the equation of motion for a random acceleration. Again, he assumed the mean step size is zero but that the random accelerations don't vanish because their dispersions remain finite. For Newton this would not have seemed at all unusual except for the assumption that the impulses are random, and perhaps not even then. But to a community used to thinking of forces as continuously acting, this treatment was strange indeed.

We only need to think of a random walk as a set of discontinuous steps. At each moment, t, we have a distribution of kicks, $f(\Delta p)$. To make this even simpler, think of this distribution as the probability of sustaining an impulse $\Delta \mathbf{p}$ that is random in both magnitude *and* direction and there are no correlations. Another way of saying this is there is no memory in the system and each impulse is strictly independent of every other. Then we're not really computing the motion of any one particle but the collection of trajectories and accelerations of the whole ensemble. Each has a separate history. In probability theory this is now called a Markov process, where the probability of being at a position $\mathbf{x} + \Delta \mathbf{x}$ at a time $t + \Delta t$ depends only on the conditions at the position $\mathbf{x}$ and not on the previous history, the step at any instant being determined by a probability distribution for the jumps in position and velocity, $P(\Delta \mathbf{x}, \Delta \mathbf{v})$, that depend on the properties of the medium. The same is true for the momentum, which constantly fluctuates around $\langle \Delta p \rangle = 0$. This independence of the impulses guarantees that the particle doesn't drift in any particular direction in space but instead always walks around a fixed location with ever increasing distance from the origin. Over the long term, the mean position remains zero (we're always free to choose an origin for the coordinate system), this can be written as $\langle \Delta x \rangle = 0$ but its distance from the point of departure—the root mean square step size, $\sqrt{\langle \Delta x^2 \rangle}$ is *not* zero. As time passes the particle executes a diffusive motion due to these microscopic impulses. On average, the smallest steps are the most probable but, every now and then, there will be the rare kick that sends the particle far from the origin. Keeping for a moment in a one dimensional picture, along a line we have as many negative as positive kicks—so the center doesn't drift—but they don't have same magnitude and that is the key for understanding this result—there is a distribution of the kicks. The probability for sustaining any kick is given by a *normal distribution*—the same one found by Maxwell for an ideal gas—if there is no correlation, and the rate is determined by the *diffusion coefficient*. Hence, $\langle \Delta x^2 \rangle = 4D\Delta t$, where the diffusion term is calculated from Langevin's equation of motion. Although this may seem only a formalism it's a result of far-reaching importance. It actually provides the means to determine the number of molecules in a liter of gas because the diffusion rate depends on the number of kicks received per interval of time, which in turn depends on the gas density. Most important, the distribution is the same as that found by Maxwell if we look at the effect on momenta of the collisions, the individual velocities spread out in the same way the particles spread out in space. The circle was now closed.

In re-introducing impulsive forcing for local motion, we have a way to determine the microscopic properties of matter by macroscopic observations. Each individual particle trajectory is unique but the *ensemble* of trajectories isn't. We may see any one particle moving unpredictably, in the sense that random interactions are constantly happening, but the motion of the collection is deterministic. The fine grained view of the world at the particle level is then replaced by the macroscopic quantities, *all of which are averages over an evolving ensemble*. A swarm of insects or a flock of birds provides the best comparison. We speak of the collective (e.g., "flock," "swarm") having a mean momentum and a total mass, but the structure of the group is continually changing as individuals interact with each other and their surrounding. There is an internal energy—each has a random as well as mean component of its motion—and any measurement can be referred to either statistically on the whole or individually on the members. For this reason, the new view provided a very powerful explanation for many of the phenomena treated by thermodynamics. If the flock is in equilibrium, that is to say the individuals are moving with respect to each other governed by some stable random process, their motions will be characterized by a single parameter: the temperature. If excited, they can disperse but maintain the same mean motion, similar to a heating. But this motion is reversible in the sense that they can reform and pass through a continuous sequence of configurations.

Chaos

But why is anything random? It doesn't seem to follow from the Newtonian principles; in fact, it can't. The classical equations for an adiabatic system are reversible in time, any trajectory can be traced back as far as we want or sent as far into the future as we desire. For instance, Laplace had imagined a *superman*, a construction from his treatise on celestial mechanics, who can see at any instant all the particles in the universe and make predictions for the motion of each with absolute certainty. But the second law of thermodynamics seems to violate this. In our flock analogy, we can reform the group spontaneously and there will be an infinite number of states—and here the phase space is *actually visible*—through which the ensemble passes. The motions are strictly reversible in the frame moving with the mean velocity of the ensemble. Yet there are cases where this isn't possible and you can picture it easily by allowing the size of the flock to increase without bound. All birds in the group look, more or less, alike and it is only our inattention that prevents us from labeling each and following it. There are cases, however, where this is actually not possible and they are surprisingly simple.

The spur to the modern theory comes from the second most elementary problem of gravitational mechanics, the motion of three orbiting bodies. To be more precise, we imagine two massive bodies orbiting around a common center of mass and introduce a third, tiny, body that orbits both. This version of the problem, called *restricted* in the sense that the third body is so small its effect on the other two can be neglected[5] and these have circular orbits. The first, analytical, solution was found by Lagrange. The gravitational potential is supplemented in the orbiting frame with a centrifugal term. There are then, in this frame, five equilibrium points, named after Lagrange. Call the two main masses M_1 and M_2 and the mass of the third body m. Then between M_1 and M_2 there is a point where the gravitational acceleration vanishes. This is the L_1 point. It is, however, clearly unstable because any tiny displacement that moves m toward either mass causes it to circulate only around that mass, albeit in a very distorted orbit. There are also two at either extreme of the principal masses, L_2 and L_3 where an outward displacement means escape. But there are two others, perpendicular to the line of centers, that are actually stable in the sense that a small displacement in the corotating frame leads to an orbit around the point. These two, symmetrically located across the line of centers, are called L_4 and L_5. Because the body has been displaced away from one of these points, the Coriolis acceleration (remember, this is *not* an inertial frame, the system as a whole rotates) produces a deflection of the body and since the centrifugal and centripetal forces are balanced results in the circulatory motion. Since the frame has an orbital frequency, there is always a helicity (circulation) in the system and the body moves accordingly under the action of this force.

Henri Poincaré (1854–1912) showed that the dynamics of systems near resonance can become chaotic. When a motion is in resonance with a driving force, when there isn't any damping, it continues to accelerate, for instance when you always kick at the same phase on a swing. The arc of the motion continues to increase until the frequency is no longer independent of the amplitude and the process saturates. When looking at the complex dynamics of the N-body problem, not only two but, for instance, the whole Solar system, Poincaré found that there is a dense set of possible resonances, an infinite number of them, and because they are so different in period and so closely spaced the system rapidly becomes unpredictable because of small kicks and other perturbations. The subject was further explored by Alexandr Lyapunov (1857–1918) in a series of studies on the dynamics of the solar system and the stability of motion near resonances.

This unpredictability also entered the microworld. In an attempt to understand entropy from a particle perspective, Boltzmann began a series of theoretical investigations of the interaction between atoms that could produce the known gas laws. Boltzmann treated the transfer of momentum between pairs of colliding particles assuming that the interactions balance. He showed the effect is to dissipate bulk, ordered motion in a sea of randomly moving particles that eventually lose their individual identities and, with a Maxwellian distribution, fill the available phase space. He identified this disorder with entropy and completed the link between Clausius' conception of microscale motions as heat. At the core of the Boltzmann program lay molecular chaos, the centrality of two body interactions and the impossibility of tracing any single particle forever. This is the same result, but for very different reasons, that Poincaré had discovered. By the beginning of the twentieth century this chaos had become a fundamental question in physical science, whether the stability of dynamical systems could be explored and how predictable are the motions over time. The study of chaotic behavior has developed into a separate, rich area. The "ergodic theorem" was an outgrowth of this molecular chaos. It states that within an adiabatic system the collisions between particles, or by extension the trajectories of an ensemble of particles moving under mutual perturbations in the manner Poincaré and Boltzmann had discovered, will pass through all possible phase space configurations if given enough time and any one of them is not privileged. The energy of the system can be defined but not the individual motions which become too "fine grained" to be distinguishable. For the concept of force this statistical view has an important effect: the measurements we make on macroscopic bodies are only ensemble averages, "course grained," so we need to compute the two body interactions and then extend them to collections. For a gas this involves molecular and atomic potentials in the electrostatic interactions and polarizations. For an ionized medium, a *plasma*, this includes the electrostatic effects of free charges. For gravity, it is best seen in the collective motions of stars in a cluster, a beautiful example of a *self-gravitating swarm*.

Applications: The Atmosphere

Although the subject of speculation since Aristotle's *Meteorologia*, the physical modeling of the atmosphere was really not begun until the nineteenth century. Two major factors contributed to this, one empirical, the other theoretical. Since we are imbedded in the environment, and the atmosphere is an enormous, multicomponent fluid enveloping the entire planet, it is extremely difficult to have a snapshot of its state at any moment. Unlike a laboratory experiment, where the scale is so small we can determine the properties of the fluid everywhere in real time to within the accuracy of our instruments, the atmosphere is simultaneously subjected to very different conditions at any moment. It's enough to remember that when it's noon at one longitude, on the opposite side of the planet it is midnight and this means one side is being heated by direct sunlight while the other is cooling. Before the eighteenth century, even the coordinate system—the relative time in different locations—was largely unknown so it was impossible to coordinate observations around the globe. Even if at one location the measurements could be made with substantial precision, for instance the humidity, wind, pressure, and temperature, it was almost impossible to connect these with another site on the planet. The second, theoretical, barrier was how to combine fluid mechanics and thermodynamics. The atmosphere is not a simple liquid: it's compressible, and this requires a very detailed treatment of both macroscopic (fluid motion) and microscopic (thermal properties, atmospheric chemistry, and the gas laws) models. Further complications come from the rotation of the planet: to solve the motion of this fluid you have to treat the problem as a fluid flowing over a spherical, rotating planet.

Three principal forces dominate its structure i.e., gravitation, pressure, and non-inertial forces. The radial (locally vertical) acceleration of gravity must be balanced by the pressure gradient in equilibrium. As Pascal demonstrated the pressure and density decrease with altitude since the the lower overlying weight of the air requires lower pressure—hence, less compression—to remain in balance. But this is a static problem. Weather, in contrast, requires a dynamical approach. What moves the air along the surface? Aristotelian reasoning served well enough for vertical motion, such as the formation of clouds and rain (light things rise, heavy things fall, air is a mixture, etc.), but what about wind?

To enter into the whole history of the subject would lead us too far astray. Instead, let's concentrate on one feature of the large scale. With the gas laws, one can understand that a disturbed parcel of air will necessarily move. It has an inertia because it's massive. But it is also a continuum, a fluid, and therefore responds to both shear and stress. Across any area, a difference in local density at the same temperature, or at different temperatures a difference in pressure, produces a reaction in the fluid. It moves with a speed that depends on the pressure difference. Vertically this isn't important because, overall, the atmospheric gas simply oscillates around a small perturbation and radiates waves that move at the sound speed. But for horizontal motions, if one region is at higher pressure than another, fluid will flow from high to low pressure. Such analogies had been remarkably successful for electrical currents. Ironically, it was much later, almost 40 years after Ohm, that the same reasoning was applied to the atmosphere. There is, however, a difference. In a circuit, or in normal small scale hydraulics, the rotation of the laboratory is irrelevant. For the atmosphere, it isn't. For slow motions, any pressure gradient is balanced by the Coriolis force. A region of low pressure, for instance, should simply be a "sink" for fluid. Produced, perhaps, by a local updraft that removes the air from a level, or by a lower density or temperature, the surrounding medium should refill the volume. The difference is that in a rotating frame no simple radial motion is possible, the Coriolis acceleration deflects the flow. Thus, a pressure gradient amplifies circulation, this is stated symbolically as $\mathrm{grad} P = 2\rho\Omega \times \mathbf{v}$ where Ω is the angular speed of the frame, ρ is the density as usual, and $\mathbf{v}$ is the linear velocity. Since the rotation is at some angle to the local plane (except at the equator where it is parallel), the velocity is immediately obtained for a steady flow. This is called the *geostropic approximation*. For the atmosphere, pressure and temperature gradients are the dominant accelerations and, for the Earth, the frame rotates so slowly the centrifugal acceleration is negligible. Without an understanding of the steady state dynamics in response to these forces, weather phenomena will appear mysterious and often counter-intuitive.

But notice that the motion is inevitably vortical. The importance of this for the interplay between the dynamical and thermal state of the atmosphere was first understood by Vilhelm Bjerknes at the end of the nineteenth century. Introduced to fluid mechanics by Helmholtz, he realized that the geostropic approach could be turned around, permitting a study of the forces driving the motions based only on the observed velocities. The spur was the circulation (vorticity) conservation theorem, discovered almost simultaneously in 1857 by Helmholtz and Kelvin. If a fluid moved inviscidly it conserves its circulation in the absence of frictional forces. Smoke rings and contrails are simple examples of the phenomenon, as a vortex slows down it expands. Conversely, a converging flow amplifies the vorticity and produces, according to our discussion, a strong pressure deficit relative to the background fluid. This is the well-known *bathtub vortex* but, as the dissipative action of ever-present viscosity—no matter how small—ultimately brings even this circulation to rest. Vilhelm Bjerknes, one the founders of modern meteorology, had studied with Helmholtz in Berlin and realized the dramatic, global consequences for vorticity in the control of cyclonic storms. His applications of dynamics to prediction of fluid motions in the atmosphere, including the fundamental notion of pressure and temperature fronts, became the basis of geophysical fluid dynamics. Now both the weather and the oceans were firmly in the domain of physical investigations.

Atmospheric motions are also governed by buoyancy. For liquids this had been extensively studied during the second half of the nineteenth century. In yet another application of Archimedian principles, a locally heated compressible fluid is buoyant relative to its surroundings. But unlike a floating body whose properties don't change with location, a moving parcel of air expands and cools if it moves into a region of lower pressure (as it will if it rises from the surface of the Earth). Depending on the rate of energy transfer between the interior and exterior of the parcel, it may be brought to rest by either loss of buoyancy (cooling relative to the background) or friction. If the background temperature gradient is sufficiently steep, the parcel will continue to rise and even if it expands will remain less dense than its surroundings. The statement that the background has an adiabatic gradient, a central concept in atmospheric physics, represents the limit of stable motion. Now combined with vorticity, this buoyancy also transports angular momentum and is subject to accelerations independent of the gravitation, hence making possible substantial vortex stretching (convergence) and amplification (pressure drops). Many of these phenomena had been studied in the laboratory and in problems studied by hydraulic engineers. They illustrate the enormous power and scope of the merger of thermal and mechanical principles by the end of the 1800s.

Applications: Stars as Cosmic Thermal Engines

The beginnings of an *astro*physics came with the merger of thermodynamics and gravitation. The development of thermodynamics and its application to cosmic bodies, other than dealing wit their motions, produced important results of cosmological significance. The most dramatic, perhaps, was the determination of the age of the Sun. In 1857, Kelvin and Helmholtz independently realized that a hot self-gravitating mass cannot stay hot forever. Since it is compressible, if the losses are slow enough that it doesn't collapse from the loss of pressure support, it must contract in a finite time because it has a finite amount of internal energy to lose. In contrast, for the Earth, this cooling can occur without any contraction: it is already solid the internal and external temperatures merely come to a balance. Since the surface is in contact with free space, the internal temperature steadily decreases and the rate of heat loss depends only on time. Assuming the initial store of heat comes only from the formation of the planet by gravitational contraction from some initial extended mass, perhaps an interstellar cloud, the rate of energy loss through the surface depends only on the temperature gradient and thus only on the internal temperature now. Fourier had described the cooling law for such a case, when the heat diffuses through the body to be radiated at the surface, and Kelvin was especially familiar with this work (having introduced the methods developed by Fourier to English mathematicians in the 1820s). However, a problem had already appeared since the energy flux measured on Earth seemed to be large enough to yield a relatively short timescale, hence age, since the formation of the planet. But the luminosity of the Sun was already well enough known that it seemed reasonable to apply similar reasoning to that body. The argument is this.

For a self-gravitating mass, by the virial theorem of Clausius we know the total energy must be a fraction of the binding energy, which is one half of the gravitational potential energy. Since the body loses energy, this potential energy much continue to decrease, and if the mass remains constant this requires a contraction. But this means the internal temperature actually continues to rise since, for an adiabatic gas the pressure and temperature increase together on compression. Thus the rate of radiative loss also increases and the contraction continues. Knowing only the rate of energy loss, L, since the total available energy is of order GM^2/R, where M is the mass and R is the *current* radius of the Sun, the timescale, t_{KH}—the Kelvin-Helmholtz timescale—is approximately $t_{KH} \sim (GM^2)/(RL)$ which is about 10 million years. This was long enough to be comfortable with human history but not with other, accumulating evidence that geological and biological evolution requires at least a factor of 100 times this interval. Another problem came from the measurement of the solar radius. It wasn't changing quickly enough. This can be done in two ways. The difficult one is to directly measure the size of the Sun over a relatively short period, years, to see if it is contracting. The other requires understanding the mechanics of the lunar orbit over a long time—you can use eclipse timing and duration to say if the Sun has changed its radius over millennia. Both yielded negative results. There was nothing fundamentally wrong with the calculation or the application of either the principals of hydrostatic balance (implying slow contraction) or thermal balance (the amount of energy lost is slow compared with the available store). The basic *scenario* was wrong. Both studies assumed no additional energy source for the luminosity. We know now that the energy for both the Earth and Sun are supplied by nuclear processes, not either simple diffusive loss of an initial quantity or the gravitational potential alone.

Nonetheless, the remarkable consequence of this application of mechanics to the stars had a dramatic effect. For the first time the irreversibility of thermal processes and the idea of force had produced a timescale. In an extension of this work, the instantaneous structure of the Sun was solved by Homer Lane in 1867, and later extended by Robert Emden in 1903, to encompass the general gas laws that an opaque body would obey. For a sphere, since the gravitational force depends only on the interior mass at any radius, you know the pressure gradient required to maintain mechanical balance throughout the interior provided you have a relation for the internal pressure as a function of density. Then the radial dependence of compression, the local density, is derivable. The only thing this depends on is the central density but the degree of concentration, the ratio of the central to average density, is given by the equation of state (the compressibility) alone. Thus measuring only the current mass (from the planetary orbits) and radius (knowing the solar distance) you can determine the central density (and therefore the central temperature). The resulting model, called the Lane-Emden equation, served through the 1930s to provide the basic model for stellar structure. A modification came from including the effects of radiation pressure, following the developments in electromagnetic theory, as a limiting equation of state for the hottest stars. But the general picture was complete with only these fundamental ideas of force and heat. With this, the science of astrophysics began.

NOTES

1. This image would much later in the century be put to extensive use by Stanley Jevons in his treatises on economy. The thermodynamic model spurred much of the early work on theoretical economics during the last third of the nineteenth century and well into the twentieth.

2. It would take us too far from our main story to describe the developments in chemistry during the following century but one thing is very important at this intersection of the two lines of development of chemistry and physics. These substances were soon found to have specific modes of interaction, and displayed specific relationships among themselves such as weight and affinities to other substances.

3. This is what we mean by independent probabilities. Supposing two quantities are independent of each other, call them X_2 and X_2 and each has a chance – a probability $P(x)$ – of having some value x. The joint chance of having a value simultaneously of $X_1 = a$ and $X_2 = b$ is then $P(\mathrm{aandb}) = P(a)P(b)$. It's the same for the distribution of independent values of the different components of the momentum; in fact, the distribution function is the same as the chance that in an interval of momentum dp you will measure a value p.

4. A degree of freedom is an independent mode of motion. Each is a way to repartition the internal and kinetic energies and can be included separately in the distribution function. Adding new modes of motion—new degrees of freedom—is possible if the particles have structure instead of being ideal and pointlike. Imagining a diatomic molecule as a harmonic oscillator that can execute one vibrational and two rotational modes increases the specific heats—two are defined, one at constant pressure (C_p), the other at constant

volume (C_v), different for the two types of thermodynamic transformations—and makes their ratio, $\gamma = C_p/C_v$, decrease.

5. Although it is common practice to call this third body "massless," it clearly can't be. Instead we mean the center of mass is between the two principal components and the third, which must have *some* mass, orbits around them.

7

FIELDS AS EVERYTHING

By no endeavor can magnet ever attract a silver churn.

—W. S. Gilbert, *Patience*

Gravitation introduced the idea of a field of force into physics, spanning the vast distances between the planets and structuring the stars. But electricity and magnetism were compellingly different and required a new way of thinking about forces. They seemed at first to require a fundamentally different treatment because they violate the notions derived from experiences with gravity. A recurring theme through this chapter will be that while gravity derives from a single property of matter, mass, and is only attractive, electric and magnetic forces depend on the composition and state of the material, are independent of the mass, can also repel, and can even overwhelm the effects of gravitation. Unlike gravity, they can be switched on and off. They produce motion but not in all bodies, seeming to contradict the third law of motion. And strangest of all, these forces seem to be linked to each other and *not* to gravity.

ELECTROSTATICS AND MAGNETOSTATICS

Imagine dividing a single bar magnet into two identical pieces so they have the same strengths at their poles and also the same weight. Take one and suspend it above the other with identical poles facing each other. Note the distance at which the two reach balance. Then increase the weight of the suspended body and note how the distance between the two decreases. Since we can use the weight as a surrogate for force, as we've been doing all along, this yields a force law. Even before the Newtonian treatment of gravity, this was enough to show that a repulsive force can have an simple, universal dependence on distance. The reverse is also simply found by taking opposite poles and reversing the position of the weights. The attractive force had provided Kepler with a physical explanation for the binding of the planets to the Sun, and it was obvious also from the orientation of the compass needle. This second phenomenon was modeled, although not explained, by Gilbert at the beginning of the seventeenth century (and exploited by Halley at the end of the same century in his magnetic map of the world) using a spherical lodestone to represent the earth and a slim filing to serve as a compass. With this, he was able to account for the navigators' results: if the Earth is a magnet, the compass traces its influence. Unlike gravity, however, magnetism isn't universal in the sense that not everything is equally affected and there are some substances that defy magnetic attraction. This *non-universality* was mimicked by electricity. Some substances could easily produce sparking and other electrical phenomena, while others yielded no effects although the phenomenon was far more general than magnetism. Electricity seems, in common experience, to come in two forms. One is the ability of a substance to retain charge and produce electrical phenomena, sparking being the most obvious effects and the one that first excited curiosity. Unlike magnetism, it can stored in the substance, can be exchanged by contact, and can be lost in time. Like magnetism, it can both attract and repel depending on the substance.

The first step was understanding the nature of this force. Since it has quantity—for instance a substance can become more charged if it is rubbed for a longer time—the "electrical fire" seemed to be something exchangeable. The phenomenon suggests an explanation. The amount of "electricity" and an object's ability to retain it depends on the the substance of which the body is composed. Unlike gravity, this isn't something intrinsic to all mass. The model that comes to mind is, therefore, a fluid, and Benjamin Franklin (1706–1790) proposed just such an explanation. Considering a single fluid, a surplus can be called *positive*, while a deficit is *negative*. Obviously, in equilibrium these cancel and produce neutrality. This single fluid picture thus explains the attraction of bodies as an attempt to neutralize by the exchange of the electrical fluid. Franklin's famous kite experiment was a demonstration of his theory, as well as of the electrical nature of lightning, and *not* a discovery. The experiment was carefully prepared according to Franklin's notion of the fluid exchange. In 1757, John Symmer, a member of the Royal Society, in one of the most amusing and profound examples of the application of a completely ordinary observation to a deep question, used the discharge of his woolen socks under different ambient conditions to argue in favor of a two fluid hypothesis, although maintaining the terms positive and negative as Franklin had used them. But without a way to quantify charge or calibrate some standards the study of electricity could not be put on the same footing as gravity. This required a way to link electrical and mechanical effects.

Inspired by the force law for gravitation, Coulomb set out to determine the law governing the interaction of charges. He faced a difficult problem in measuring the force between two charged bodies. Repulsion is comparatively easy because there is no restraint on the charged objects. They can't touch. So for this measurement he could use the torsion balance. Two small spheres were identically charged, using a pin to transfer charge from a Leyden jar to a pair of pith balls. These were mounted on the arm of a torsion balance for which Coulomb had already performed the appropriate calibrations. Thus the static twist of the cord of the balance gave the measure of the force. For the opposite case, attraction, he charged a metallic sphere and used induced charge on a grounded sphere to produce the opposite sign. The spheres had, however, to be kept separated so an alternate means had to be found for measuring the force. Coulomb had the idea of performing a *dynamical* measurement. Perturbing the equilibrium position, maintained by tension, he determined the oscillation frequency, ω, of the horizontal pendulum. The frequency scales with a force F as $F \sim \omega^2$. If the force between the attractive spheres scales as the inverse square of the distance, as he had found for repulsive charges, then the time of oscillation Δt should scale as $\Delta t \sim d$. The same types of measurements were done for magnetics with the same result: the force between two objects either electrified or magnetized varies as the inverse square of the separation to very high accuracy (about 2% in his measurements). This was the technique later used by Cavendish to measure the gravitational constant but it is even more striking with electricity because this force was not part of the original mechanical worldview.

ELECTRICAL POTENTIALS

When we discussed Green's theory of potentials we focused on gravitation. But the route by which he arrived at his construct was actually the study of electricity and charge. He worked within a fluid conceptualization of the force, an atmosphere that surrounds and permeates a conductor. This electrical fluid can change density depending on the properties of the medium but has certain boundary conditions. Just as there must be pressure balance across an equipotential surface for a fluid, there must also be equilibrium between the electrical fluid within and outside a conducting medium. His principal mathematical results were reached with electricity in mind. He applied the condition that the fluid density, or potential function, may be discontinuous across the boundary of a conductor. He also stated that the component of the force perpendicular to the surface may be discontinuous but this is balanced by the development of a surface charge. The potential function then depends on the continuous distribution of the charge with each volume element contributing to the potential weighted by its distance. Thus Green was able to recover Poisson's result and also match the solutions inside and outside a conductor. For gravitation hollow spheres are irrelevant but for electrical and magnetic phenomena these are shells that store surface charge. It was therefore possible to compute the capacity of any shape to sustain a given potential difference and, therefore, explain the properties of condensers.

FROM ELECTROMAGNETISM TO ELECTRODYNAMICS

Electrical and magnetic phenomena had always been studied separately although they share many features. The most striking is polarity, an obvious term for a magnet. Furthermore, until the start of the nineteenth century, magnetic phenomena had been mainly investigated on their own using permanent magnets. When, in 1820, Hans Christian Oersted (1777–1851) discovered the effect of a circuit on a nearby compass needle it was not entirely by chance. He had in mind the polarization of the battery and the possibility that the fluid, coursing through the wire, might exert a force on a polar body. His discovery had particular drama because the actual first observation of the action of the circuit occurred during a lecture demonstration in which he failed at first to see any deflection when a compass needle was placed perpendicular to the wire. But when the needle was placed nearly parallel to the axis of the current, it oriented itself perpendicularly after first oscillating. In the early nineteenth century, even rather powerfully charged batteries had a tendency to discharge quickly so the confirmation of the effect and its exploration took some time. But by the end of the year he had also discovered that the effect diminishes with distance from the axis and, more significantly, that by displacing the needle around the wire as Gilbert had around the terrela, Oersted found that the effect was circumferential. It was as if a vortex surrounded the wire and oriented the magnet much as a flow would orient a piece of wood. Yet this was outside of the current and away from the source. The effect vanished when the battery was turned off or spent.[1] The news of the discovery spread rapidly among the scientific societies of Europe where the demonstration was eagerly repeated. It arrived in Paris with Biot. One in the audience, André-Marie Ampère (1775–1836), was to become, in the words of a later master figure in the study of electromagnetism, Maxwell, "the Newton of electricity." Within a matter of weeks his first communication extending and amplifying Oersted's results was received and with it a new physical picture was formed.

Ampère's most important result was the discovery of mutual influence of identically and oppositely directed currents. Since the battery is polar, reversing the poles changes the direction of the flow. Thus it was that Ampère discovered that suspended wires carrying oppositely directed currents repel while if passing in the same direction they attract, precisely like permanent magnets and speculated that the magnetism of the Earth arises from some sort of current. But the magnetic field of the Earth had been explained by the flow of some subtle fluid that entered at one pole and exited at the other, much like Franklin's earliest conception of electricity. This showed, instead, that the magnetic might themselves harbor internal currents. To explore this, Ampère reasoned that a change in the way the current is moving through the wire would change the external field and, in particular, if the flow were helical the resultant field would be axial and polar. Now, it seemed, the nature of this magnetic force was seated, somehow, in the electrical action of the battery and current. The stronger the action, the stronger the created field. Ampère coined the term *electrodynamics* to describe the whole range of physical effects he observed. His approach was to imagine current elements, analogous to charges, along the wire. Rather than explaining the origin of the orientation of the field, he asserted it. In a similar manner to the dependence of the gravitational force between two bodies on the product of the masses and of the electrostatic force between two charges on the product of their charges, he asserted that the magnetic force depends on the product of the two currents. But since these are perpendicular to the direction of the force, the description is fundamentally different than the radial one for the other two, static fields.

A new force had been discovered, *electromagnetism*. It changed how to describe the source of a force from static charges to dynamical currents. But the effects were still described in the same way as Newtonian gravity, a sort of action at a distance. There is some quantity of a "stuff," charge or magnetic fluence, that remains constant in time. Once the charge, for instance, is put into a body, it induces a force and causes motion, but it doesn't require something moving to create it. On the other hand, Oersted's experiment showed that something had been actively passing through the wire of the circuit to produce a magnetic effect. Stranger yet, instead of acting along the line of centers, the force caused a torque of the compass needle and acted at right angles. The Biot-Savart law for the magnetic field produced by a current and Ampère's law for interaction of circuits and Coulomb's law for the force between static charges are all similar to Newton's gravitational law, a force varies dimensionally as the inverse square of the distance from the source. The essential difference is that each element of one circuit acts on every element of the other at right angles and depending on the inverse cube of the distance, the force between two current elements depending on their product.

Another difference between gravity and the electromagnetic fields was quickly noted: a change is induced within the body by the presence of the field, it polarizes. For a magnet, this causes the poles to torque and anti-align. For electricity, a static charge image develops on the surface of the affected object, an opposite charge, when the field is applied. For gravity, the differential acceleration of the tides might look like this since, in the orbiting frame, the accelerations are equal and oppositely directed. But in reality, in the stationary frame, the force is always attractive and centripetal and it is only because of the finite size of the body that the force varies across it. The mass doesn't polarize although the distortion aligns. Nothing actually changes in the interior and the reaction, while depending on the rigidity of the body, is otherwise independent of its composition.

ACTION-AT-A-DISTANCE

Gravitation was somehow communicated instantaneously between masses. In the *Principia*, no explanation was proposed—it was one of the "hypotheses" Newton sought to avoid—but there was no pressing need for such an explanation. The distances could be handled without difficulty because there were two features implicit in his thinking that permitted this. One was that there was no specific timescale for any dynamical processes. If the propagation of the attractive influences of masses took time, that was fine because one had all the time in the world, literally. The other was the action of God, who was always there to adjust the workings of anything that got out of balance. This second was acceptable in the seventeenth century way of looking at natural philosophy and, actually, was a plus for the picture because it squared nicely with the theological preconception of finding in the Universe the action of a watchful creator. The former was more troubling because it used the wrong language, the sort of semi-mystical description that irked the Continental philosophers and smacked of alchemy and the occult.

For electricity it was the same despite the anomalous feature of repulsion. But electrical and magnetic phenomena are not only both attractive and repulsive, they are also polar. For a magnet this is obvious, in fact it's part of the definition of magnetism. For electricity, there were two competing views in the eighteenth century. Franklin's single fluid hypothesis used "positive" and "negative" in the sense of "excess" and "deficit." If the equilibrium state is neutral, an excess of charge displaces to cancel a deficit when the two are brought into contact. An insulator prevents this transfer while a conductor facilitates it. This squared well with experiments. But it also worked well in the fluid context. For a fluid column in equilibrium under gravity, there is no net acceleration although there is an internal pressure that counterbalances the weight of the fluid at every point. If, however, a hole is opened in the container, a jet of fluid exits and reaches the speed it would have were it to be freely falling from the top of the column to the hole. This can be demonstrated by noting the height reached by the jet that is tilted upward and also its range. Thus Torricelli's law of pressure, combined with the picture of an ideal fluid by Euler and Daniel Bernoulli, described the motion. The imbalance of forces at the hole, when the fluid pressure exceeds that of the surrounding atmosphere, accelerates the fluid. There was a dispute for some time over the nature of this "electrical fluid," whether there is one that is in excess or deficit, or two that are truly attractive and/or repulsive.

The first attempts to determine the force between charges was made by Joseph Priestly (1733–1804) who also demonstrated that within a conductor there is no net force. This was one of the foundations on which Green's potential theory was built. When Coulomb performed his electrical measurements, finding that the force between static charges depends only on the inverse square of the distance, it seemed a confirmation of the universal picture of forces and action-at-a-distance that Newton had propounded. In fact, it was precisely because Coulomb was dealing with a static effect that this was plausible. The measurement was made using a torsion balance, adopting the technique also independently proposed by Mitchell for measuring the gravitational attraction between two masses and later adopted by Cavendish (as we will discuss shortly). In addition, Coulomb demonstrated the existence of two separate charges that, when alike, repel and otherwise attract.

When it came to electrodynamics, specifically Oersted's and Ampère's discoveries, this started to present a conceptual problem. Ampère could treat the effects of circuit elements as if they were charges, little pieces of lines in closed circuits that in the presence of a steady state could act for a long enough time that propagation effects were negligible. The Biot-Savart law and Ampère's law were both couched in this language, with the magnetic field being proportional to the current in a wire and the force between two wires varying dimensionally as the inverse square of their separation and the product of their currents. But open circuits were a different matter. First we need to distinguish the two cases. A *closed* circuit is simple, for instance a loop of wire connecting the poles of a Voltic pile. A flow of electricity, whatever that might be, within this "pipe" creates a magnetic field and that is the cause of the repulsion or attraction when two such circuits are brought into counter-alignment or alignment, respectively. An *open* circuit, in contrast, is a discharge, and this phenomenon also differentiates electrical and gravitational phenomena. It is possible to collect charge—in the language of the eighteenth century, to "condense the electrical fluid"—for later use; the device, called a Leyden jar. Franklin demonstrated that this effect can be multiplied by combining the jars in series, he called it a "battery of condensers," and showed that the stored charge is held within the insulator (later Priestly showed the glass can be replaced by air). There is a finite quantity of charge that can be stored, depending on the insulating substance, again unlike gravitation. The charge effect resides at the surface of the insulator, the part that is either in contact with the metallic shield or at the gap when they are not in contact. When the Leyden jar releases its charge it obviously takes time. This too is completely different from gravity. Since the source isn't in steady state, its effects on anything in the vicinity should also depend on time and to require propagation of the influence. Static approximations could not explain this effect. Several laws were proposed that included the velocities and acceleration of the elements of the electrical fluid and/or the circuit but none of these were completely specified nor epistemologically consistent with the prevailing physical theoretical picture.

Circuits as Flows

A static fluid representation had served well for potential theory. It allowed a simple mechanical image of how a charge medium could accommodate itself to the presence of a central, electrified body. The discovery of electromagnetism, the connection between a time dependent electric potential and the generation of a magnetic field, suggested an extension to a dynamical fluid. This was accomplished by introducing currents. The model was remarkably successful. You'll see shortly that even Faraday's force lines could be expressed mathematically as results of currents generating vortical motions of an imbedding continuous medium that transmit the electrical influence to magnetic bodies. In this way, the Biot-Savart law and Ampère's treatment of sources were accommodated in a single explanatory picture. It wasn't only fruitful for explaining the generation of magnetic activity by a discharging pile.

When a wire connects the poles of a battery, a current flows from one site to another. In the single component picture, this is the transfer of the "electrical fluid" from that which is in excess to the deficient side. It was clear that this can occur in free space, or in air. To have it pass in a closed circuit meant that the charge was moving as if in a pipe driven by an *electromotive force*, the *EMF*. The hydrodynamic analogy of currents, sources, and sinks was effectively employed by Georg Simon Ohm (1789–1854) to study the passage of currents through circuits and their resistive elements. For us, this is only a sideline but we still have reminders of this earlier era in the terminology of physics and electrical engineering. The EMF, which is now called the *potential difference* or *voltage*, records the confusion between force and energy that persisted through the nineteenth century. In an electrical circuit, the EMF is the source of the current, the battery or power supply, and a current flows in response. The hydraulic example would be water issuing from a container, the electrical analogy is a condenser, placed higher than the sink so a current flows through the tube and then using a pump to raise the water again, the analog of a battery. This immediately connects the mechanical and electrical forces, a potential difference is needed to create a flow and maintain it. The introduction of resistances to the flow can occur in series or parallel, you can think of this as joints in the piping or junk clogging the tube, thus impeding the flow, and the two laws that result are that the sum of all currents is constant and the sum of all drops in the potential must equal the potential difference of the source. For our discussion, this amounts to a new way of seeing how a force is *transmitted* through a system, how various components act as modifiers and amplifiers. The capacitor, resistor, and inductor were all *loads* that impeded the flow of the current. A condenser was not only, as with a Leyden jar, something with the capacity to store charge. It became a dynamical element of a circuit, capable of both charging and discharging in time. This general fluid conception, first subjected to systematic mathematical treatment by Ohm, also immediately connected the *electromotive force* with the energy, or potential, of the pile. The general law is that the potential difference across a circuit elements is equal to the current times the resistance. Further use of the flow analogy was made by Gustav Robert Kirchhoff (1824–1887) in his analysis of how loads distribute the current and potential difference. Loads placed in the circuit in series, that is in immediate succession, are directly additive. This is because the total drop in the potential across the sum of the elements is the same as between the poles of the pile. The same quantity of current must pass through each element, as would happen in an incompressible fluid. Therefore, $Z_{\text{total}} = Z_1 + Z_2 + \cdots$, where Z_j is the impedance (the generalized load). On the other hand, as with a cascade, parallel elements experience identical potential drops but pass different amounts of current. Think of the different parts of Niagara Falls: the height from which the river drops is the same for all channels but they pass different quantities of water. The potential difference is now, however, not due to gravity but electric field that develops across each elements. Thus, for parallel loads the impedances add in inverse series. $1/Z_{\text{total}}) = (1/Z_1 + 1/Z_2 + \cdots$. The reigning confusion in terminology between "force" and "energy" persisted, electromotive "force" (still so-called) is actually the potential difference across the circuit.

Electrical phenomena, especially those of circuits, also displayed links to thermodynamics. Resistances become hot, wires glow, and the voltic pile gets hotter as it operates. For instance, while the temperature is defined only for a thermal system, the pressure and volume can be directly relate to dynamical and geometric quantities. Since the work done by the system is related to the temperature, T, which is related to the internal energy. This is precisely what happens in a battery, a chemical process releases energy and, somehow, also generates a potential difference that does the work of driving a current. A scent of unification was in the air. This connection between work and mechanical reaction was the central feature of Helmholtz's memoir *On the Conservation of Forces* in 1847, which I introduced when discussing frictional forces and dissipation. It was the first precise treatment of energy. Although it later became the standard survey of the subject and is now considered a classic, its significance was not immediately appreciated. Helmholtz has serious difficulties getting it published since it was his first major contribution in physics written while still in his twenties.[2] You'll notice the title. Even at this comparatively late stage in the development of mechanics there remained an ambiguity between the words "force" and "energy" but the memoir makes the reason obvious. Helmholtz started with the dynamical equations and showed that for a central potential field, the work plus the *vis viva*, the kinetic energy, is constant. He then showed how to include friction, the loss of *vis viva*, in this sum by introducing the concept of internal energy, which he could then link to heat and temperature. Perhaps the most strikingly original part was, however, the connection he drew between the voltic pile—current in electrical circuits—and energy. Helmholtz proved that the rate of power—the rate of energy dissipation—for a resistor is related to the electrical potential driving the current and the resistance, much as viscosity in a flow slows the motion. As I've said, a notable phenomenon in circuits is that the components heat up. This connection between thermodynamics and electrodynamics seemed like an effect of friction, almost a viscosity of the current in the circuit. His result was that the power $= I^2R$ for a current I and resistance R. He already knew, from the definition of capacitance, that the charge is related to the potential difference and since the charge times the EMF is the energy (analogous to the mass times the potential), the current times the EMF is the power (the rate of change of energy with time). It was thus possible to extend all previously developed fluid mechanical experience to electrodynamic phenomena and account for gains and losses. The conceptual framework was very general and individual devices could be included in the same way as Kirchhoff's treatment of the impedances.

Figure 7.1: Hermann von Helmholtz. Image copyright History of Science Collections, University of Oklahoma Libraries.

Induction and Lines of Force

Charge also induces charge; when near a neutral material a charged body can produce an electrical response. Charge can be transferred from one material to another by contact and stored in some and not in others. A glass rod can hold charge and produce a polarization and conductors behave as if they have equal and opposite charges when placed near a source, charge images that are the analogs of optical mirrors. Such static effects were well studied by the 1830s, although not understood. To add the confusion, when Joseph Henry (1797–1878) and Michael Faraday (1791–1867) discovered electromagnetic induction and self-induction they introduced a new wrinkle into the concept of force, that one agent in the interaction could induce a similar state in a previously undisturbed body and even have a back-reaction on itself. This electromotive force was different from that driving a closed circuit with a battery but it required a closed loop and a time variable source. Both also discovered *self-inductance*, the back reaction of a time dependent circuit on itself, but it was Faraday who recognized the deeper implications of the discoveries and profoundly altered the course of physics.

For gravitation, the presence of another mass produces a reaction but doesn't induce anything. The force depends only on the quantity of matter. The same is true for an electric field, as we've seen, the main difference being the possibility

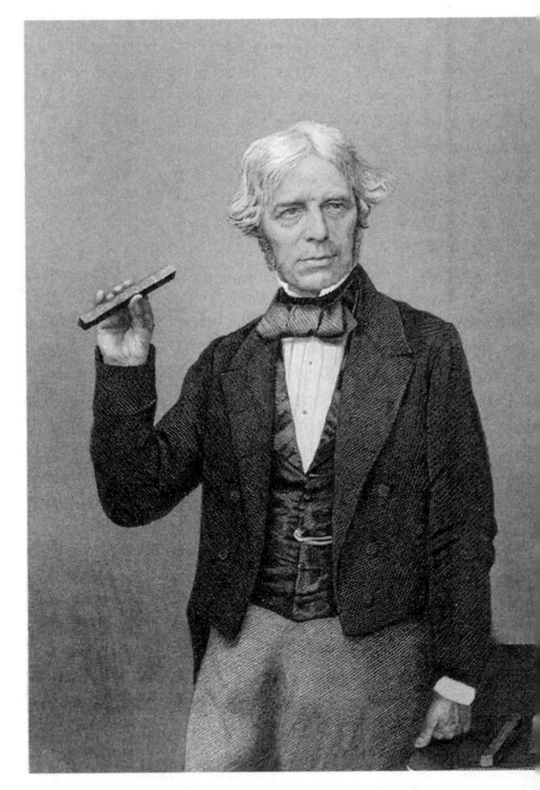

Figure 7.2: Michael Faraday. Image copyright History of Science Collections, University of Oklahoma Libraries.

of static repulsion. But both forces are radial and steady, if the source remains constant. For electromagnetism, induction is something completely different. The time dependence was the clue that Faraday used to develop a working hypothesis of enormous reach: lines of force. Placing iron filings on a sheet of paper lying on top of a permanent magnet, the filings delineate the curving lines of force that permeate space around the source. The same visualization technique, applied to a loop of wire, revealed what in the surrounding space what had already been traced out by the motion of a compass needle, that the curved lines look like a vortex around the source—the current carrying wire—and have the same properties as those of a magnetic. The visualization is now so familiar, indeed it's now a standard demonstration, it may be hard to imagine when it was new but everyone who sees it is entranced by the beauty of the effect. The discovery was announced in 1851 by Faraday in the 28th series of his *Experimental Researches in Electricity and Magnetism* in which he presented a new model for how a force is communicated by a magnetic field and thus established a novel and far-reaching program for the study of fields in general. He grounded the description in operational terms—how do you measure the direction and strength of the field—but to do this he needed to use the response of a magnetic needle and, therefore, instead of calling this topological "thing" a field, he introduces the term *lines of force*. His description is beautifully clear and I'll reproduce it in full here.

> (3071) A Line of magnetic force may be defined as the line with is described by a very small magnetic needle, when it is moved in either direction correspondent to its length, that the needle is constantly a tangent to the line of motion; or it is the line along with, if a transverse wire be moved in either direction, there is no tendency to the formation of any current in the wire, whilst if it is moved in any other direction there is such a tendency; or it is that line which coincides with the direction of the magnetcrystallic axis of a crystal of bismuth, which is carried in either direction along with it. The direction of these lines about and amongst magnetics and electric currents, is easily represented and understood, in a general manner, by the ordinary use of iron filings. (3072) These lines have not merely determine direction, recognizable as above (3071), but because they are related to a polar or antithetical power, have opposite qualities or conditions in opposite

> directions; these qualities which have to be distinguished and identified, are made manifest to us, either by the position of the ends of the magnetic needle, or by the direction of the current induced in moving the wire. (3073) The point equally important in the definition of these lines is that they represent a determinate and unchanging amount of force. Though, therefore, their forms, as they exist between two or more centers or sources of magnetic power, may vary very greatly, and also the space through which they may be traced, yet the sum of power contained in any one section of a given portion of the lines is exactly equal to the sum of power in any other section of the same lines, however altered in form, or however convergent or divergent they may be at the second place.

Faraday's language here is precise and geometric. Although not explicitly expressed in symbols, it leads to the differential equations for the lines of constant force; in fact, the words translate virtually directly into symbols such as divergence and line sections. As he states, these can be easily visualized by sprinkling iron filings on a sheet of paper above a bar magnet, although Faraday doesn't note that this is actually visualizing the field in three dimensions since the filings near the poles will stand perpendicular to the paper due to the convergence of the field. You might be thinking that the idea of field lines, indeed of field in general, could be read into Gilbert's description of the terrela. Indeed, it often is, but there are several important differences. Faraday's experiments contained a basic ingredient not found in the early descriptions of the environment around a magnetic: for him, there was an intrinsic connection between the fields *within and outside of* the magnet. The effects are not due to emanations, humors, or emissions, they are probing a *field*, a static space filling force. When these lines are changed at any region of space, that is when the field is altered, there are physical effects. Instead of being static entity, the field is active and it can induce time dependent reactions. It is truly a *force* in the sense that it can produce an acceleration.

When Faraday discovered the time dependent effect, that a change in the field produces a current, he saw in this a generalization of the force line idea. If the *number of lines of force* passing through a closed circuit element, such as a wire loop, changes with time, it produces an electromotive force that induces a current through the wire in the absence of a permanent driver (e.g., a battery). The faster the change, the greater the induced current, therefore the greater the EMF. This effect could not be instantaneous because it depended on the rate of change, with time, of the flux passing through a closed area and it was the first electrical or magnetic effect that could not be produced by a static field, as Faraday also demonstrated. Importantly it didn't matter whether the magnetic or the circuit moved, thus showing that the cutting of the field lines is completely relative. Thus, with a single observation, the concept of force permanently changed. It was no longer possible to ignore completely the propagation of the effect of the source. Inserting a soft iron rod in an electrified, current carrying coil produced a magnetic field; winding another wire around some other free part of the rod reacted with an induced current that depended on the rate of change of the driving current. This is a "transformer." With this field model Faraday could describe a broad range of new phenomena. For instance, if a loop of wire is moved so it cuts these field lines, it induces an electrical current. What's required is to change the field passing through the circuit in time. This can happen either because the strength of the field *is* changing, as demonstrated later by Lenz with a moving magnet passed through a coil that generates a current to induce an oppositely directed field, or by the relative motion of the wire and the field as in the dynamo. The application came rapidly: a generator of electricity that could convert motion into current and, in turn through the action of electric fields, produce work.

The Dynamical Basis of Electromagnetic Units

How do you standardize the units of electromagnetism? Within mechanics, mass (M), length (L), and time (T) are the basic quantities and because of the force law can be used to define all dynamically related quantities. For instance, the unit of force (the *dyne* or *newton*, depending on whether cgs or MKS units are employed) can be written dimensionally as MLT^{-2} (that is, mass times acceleration). Any stress is the force per unit area, so whether pressure or compressional stress this is $ML^{-1}T^{-2}$. Energy is ML^2T^{-2}, and so on. When we're dealing with electrical or magnetic fields, this isn't so obvious. How do you define the unit of electrical potential? Force is the key: the potential for a gravitational field has the same dynamical role for mass that electromotive force—potential difference—has for electricity. It produces a work and thus has the units of energy per unit charge. Since the electrical interaction between point masses is a force, dimensionally $Q^2L^{-2} \sim MLT^{-2}$, where Q is the charge, producing the unit for charge $M^{1/2}L^{3/2}T^{-1}$. Recall we used the same argument for the gravitational interaction to find, using the elementary units, the dimensions of the gravitational constant. To extend this to any other field components is relatively simple. Since potential differences produce work, the units must be the same as energy so that $QV \sim ML^2T^{-2}$ which, when we have defined the potential difference V as a line integral over the electric field provides a definition of the electric field $E \sim VL^{-1}$. Further, defining capacitance as the charge retained by a material acted on by some potential, these units are extended to indirect properties of charged systems. The point here is that without the third law, it is impossible to consistently use elementary dynamical principles to relate the causes of the phenomena and their magnitudes (the fields) to their effects (e.g., currents in circuits). For magnetism we can continue along the same lines using the inductive law. The galvanometer and ammeter depend on the electromagnetic interaction between coils and fields, through currents. Static magnetic fields are more easily arranged using naturally occurring materials (e.g., lodestones) whose effects can be measured. But for standards, the units for the fields depend on the electromagnetic principles. Because only the static electric field is defined dynamically, without the induction law, hence the dynamo, it isn't possible to obtain a set of dimensioned quantities to represent the magnetic field.[3]

MAXWELL: THE UNIFIED THEORY

Electrodynamical effects, especially induction, cried out for a unification. The inter-convertibility of the actions of the two fields, depended on both space and time and, although was fundamentally different from gravitation, nevertheless could be understood mechanically with the new conception of field that Faraday introduced.

At this stage, with the field concept still being debated, we come to the work of James Clerk Maxwell, who has already appeared as one of the founders of statistical mechanics. His successful combination of electricity and magnetism within a single dynamical theory represents the first unification of the fundamental physical forces and established the research program that is still pursued today. Although Newton and later researchers extended mechanics, the two developments—researches in electrical and magnetic phenomena and those of mechanical systems—remained disjoint. Helmholtz, Weber, and Kelvin, in particular, contributed many elements to the developing picture. But the unification was accomplished by Maxwell alone, although the clarification of the nascent theory was necessarily carried on by others than its author, who died at age 48 of abdominal cancer.

Figure 7.3: James Clerk Maxwell. Image copyright History of Science Collections, University of Oklahoma Libraries.

Maxwell's innovation came from taking the fields as properties of the medium. The field outside a dielectric, for instance the field in free space between the plates of a condenser, is the electric field **E**. A dielectric, insulating material inserted between the plates can increase the charge storage capacity of the condenser. This is because it polarizes, developing a surface charge produced by **E** and the field *inside* the dielectric is **D**. Because Maxwell viewed this as a response to the stressing of the dielectric by **E** he called it the *displacement*; it was an electrical analogy to the strain of an elastic medium. By a further analogy, the external field produced by a magnetic is **B** and that inside a paramagnetic substance is **H**. The electrical polarization, **P**, alters the electric field within the dielectric, the magnetic polarization **M** does the same for the magnetic field. This led to the pair of definitions, $\mathbf{D} = \mathbf{E} + \mathbf{P}$ and $\mathbf{H} = \mathbf{B} + \mathbf{M}$. Both **P** and M are induced and depend on the external field strengths. The field in the medium, if stationary, is produced by the polarization of the charge, this was already known from the behavior of dielectric materials that reacted to an external field by displaying a surface charge. The change in the field between the interior and exterior, as envisioned first by Green, could now account for the surface charge in any medium with a finite resistivity. The field within the medium is then given by **D** and the internal field is reduced by the production of a polarization **P** which produces a surface charge σ. Thus the difference between the internal and external displacement current is the origin of the surface field and the divergence of the electric field is given by the presence of a net charge density. Maxwell could then write two proportions that depend on the composition of the material, $\mathbf{D} = \epsilon\mathbf{E}$ and $\mathbf{B} = \mu\mathbf{H}$ where ϵ is the dielectric constant and μ is the magnetic permeability. Neither of these was known in an absolute sense but could be scaled to the field in "free space," so the limiting values of the two constants are ϵ_0 and μ_0, respectively.

But what about *dynamics*? Maxwell knew from Coulomb and Cavendish[4] that a vacuum field produced by a single point charge that is, it produces a flux whose intensity falls, as for gravity, as the inverse square of the distance. But he also knew that within a medium, the divergence of the field strength is produced by a change in the density of the charge. This was Poisson's, Green's, and Gauss' result. Any change in the charge distribution with time is equivalent to a current, but by the continuity condition it must be balanced by an influx or outflow of current. Since he could write an equation for the change in the field in time using the Poisson equation to link the divergence of the internal electric field, **D** to the density, the rate of change with time of the density was the same as the change in time of the divergence of the field. Maxwell then distinguished two types of currents. One is the free current, **J**, a charged fluid for which an imbalance of the rates of inflow and outflow through some volume produces a change within the volume of the density of free charge. This is the continuity condition and is precisely analogous to the flow of a compressible fluid. This free current requires a driving agent, the electromotive force or potential difference across a circuit. But continuing the analogy with an elastic medium, Maxwell introduced the *displacement current*. He reasoned that there is some kind of strain on the material produced by the stress of the applied field so if the field changes in time so should the displacement. The orientation of the electrical units changes but, being bound to each other and rigidly situated in the material, the charges cannot move freely. Instead, they reduce the internal field while accumulating at—or toward—the surface of the body, producing a time dependent reaction and a variable polarization when any changes occur in the imposed electric field. The effect had already been quantified by Faraday and Mossotti had linked it to the index of refraction of a medium. The same reasoning applied to the magnetic field. We have a link between a fluid and elastic picture for the reaction of any body. The mass, you notice, doesn't matter.[5]

The electric field can be obtained from a potential, which is a scalar quantity, whose directional changes in space—its gradient—gives both the strength and direction of **E**. The magnetic field, on the other hand, is a *polar* quantity (that is, after all, one of its main properties, it has a handedness) so it must be due to a different kind of potential function. To find it, there are several clues. One is Ampère's law. Since a current is required to produce a magnetic field, the potential must be a vector (remember, this is the current density **J** and not simply the number of charges per unit time). Certainly some sort of spatial derivative is needed as well to produce a field but, in this case, it must have polarity. The mathematical representation was known from fluids, it's the same one required to describe vorticity and Maxwell therefore adopted a circulatory potential function **A**. The two fields must be unified, eventually, because the current connects these through the induction law since the motion of a current through a field induces an electrical potential difference. Then the current serves as the source term for the vector field just as the static charge is the source of the electric field. The Biot-Savart law states that a current is necessary to produce a magnetic field in space and, following Oersted's experiment, that this field acts between circuits perpendicular to their individual elements. But at this stage the treatment becomes truly novel: *even without sources, a time dependent strain in the medium acts like a current.* For Maxwell, this meant that the medium—the ether—if subjected to some forcing, would propagate that disturbance by itself.

To sum up, Maxwell required four fields, instead of two, with different sources and different constraints. The internal fields are the displacement and the magnetic flux. The external fields are the electric field strength, and the magnetic field strength. The connection between these depends on the properties of the medium. For the magnetic field, this is the susceptibility while for the electric field this depends on the dielectric constant. Both of these are measurable and linked to the energy of the field contained in an arbitrary volume, V, which he found is the integral Energy $= \frac{1}{2}\int[\mathbf{E}\cdot\mathbf{D}+\mathbf{H}\cdot\mathbf{B}]dV$. The first term was already known from Faraday's investigations, that the energy stored in a condenser is the work required to move the charge by some displacement. If we take the displacement, **D**, literally, then the work in a volume element is the force density applied along the direction of displacement, $\mathbf{E}\cdot D$ and similarly for the magnetic field.

THE ETHER

Newton was able to avoid openly describing the *origin* of gravity, leaving it aside as a metaphysical problem (although not without much private speculation on the topic). It was enough that the force was shown to be identical for all matter and with this to explain a vast swath of physical phenomena. Even when the field construct was introduced along with its representation by a potential from which forces arise, it was still possible to avoid asking questions about the material properties of this medium. You could have action at a distance without detailing how it happens. There was no such luxury when it came to electromagnetism. The discovery of induction showed that time is fundamental to its action. The flow of charges, a current, produces a magnetic field and the rate of change of a magnetic field produces a current. Optical phenomena come from the transmission of a time dependent field between a source and receptor and this seems to require a medium. If the currents behave like fluids, then waves in fields should behave like waves in fluids with tension, inertia, and elastic properties. It seemed a return of a discontinued item from the catalog of models, a space filling substance that transmits force. This time it wasn't something as fuzzy as the universal Cartesian substance but it bore striking resemblances. This hypothetical electromagnetically active medium required vortices for magnetic currents and the ability to support wave transmission so some sort of elasticity. By the boundary conditions for electric and magnetic substances, the limiting condition for a space containing no free charges—a vacuum from the point of view of matter—still required some displacement current. Just as the index of refraction for light being defined as unity in the limit of free space, the dielectric constant and permeability of the ether were constants "in the limit"; a vacuum state is simply one that lacks free charges but can otherwise be completely filled. The same was true for gravitation but the Newtonian concept required an impulsive transmission of the force and otherwise eschewed any supporting medium as being unnecessary.

To explain induction, Maxwell translated Faraday's notion of force lines into a local setting, stating that the change in the magnetic flux passing through a circuit induces a potential difference. The integral expression of this law has a striking resemblance to vortex motion, a point not lost on either Kelvin or Maxwell. If we define the potential difference as a sum over parts of a circuit element through whose area passes a time variable magnetic flux, then the electromotive force is given by the change in time of the magnetic flux passing through a closed circuit or loop. Kelvin and Helmholtz had used a similar appearing definition for circulation (arriving at the theorem that in inviscid fluids this quantity is constant). The vortex analogy was strengthened by the condition that the magnetic flux cannot diverge. Only circulation is possible for the "fluid" supporting the field, since a vortex must either close on itself or start and end only at the boundaries of the medium. In a boundless medium, as the ether was thought to be, the second option makes no sense so the field must be produced by *dipoles* or even multipoles of higher order. The system is now closed. One set of equations describes the conservation constraints, those are the divergence conditions. The other pair are the evolution equations for the fields as they couple. A potential difference is created between any two points in space merely by changing the magnetic flux passing through any area. Consequently, a force depends only on a local variation of a field that may have a very distant origin. The other is that even absent free current, and this was Maxwell's condition for a *vacuum state*, the magnetic field could be changed locally by the straining of the ether. To be more precise, if the ether is space-filling there is no real vacuum—empty volume—anywhere. But it will seem to be empty in the sense that no matter exists to support, by its conduction of a free current, a local source for the field. In the limit that the free current vanishes, two linearly coupled fields persist, (**E**, **B**) that because of the divergenceless property of magnetism propagate as a transverse vibration of the ether. This solution, which depends on the material properties of the ether in Maxwell's theory, combine to form a wave equation. Thus the unification of the forces governing the motion of charged particles produces a vibratory mode of energy transport that resembles light. It was this result, so sweeping in its conclusion, that was the fundamental breakthrough of the theory. If only currents within the fluid were allowed, no waves would result since because of the shearing of the electric field, the divergence of the free current vanishes and the density of charges decouples from the problem. But because of the strain term for the displacement current, there is always a time dependence produced by the imposition of a time variable field and the system closes without invoking the electrostatic limit. Even polarization of light could be explained because the vibrations occur in a plane perpendicular to the direction of propagation of the wavefront. Circular polarization was just a change of phase of the two electric field components.

You can see how this enters from the equations for the two field strengths **E** and **H** fields in the absence of free charges. The constraint equations for the magnetic flux and the displacement, their divergences, both vanish (this implies that the medium is incompressible) and the time-dependent equations for the two fields combine in a single equation. The argument proceeds this way. The spatial variation of the magnetic field, even when no free current is present, is produced by the time dependent differential strain—the displacement current—while the spatial changes of the electric field are produced by time variations of the magnetic flux (or induction). The two resulting equations are linear and the two components of the electromagnetic field don't interfere with each other. The time and space dependent equation that resulted for *each* field separately is a wave equation, the mathematics of which had been studied for over a century since D'Alembert and Euler. Maxwell immediately recognized it and, further, identified the propagation speed with the product of the two material constants that are the analogs of mechanical strains, ϵ and μ. This *dynamical* field now propagates through the ether with a constant speed that depends only on the electromagnetic properties of the medium. While forces, as such, are not mentioned here, we are clearly dealing with a mechanical problem in which a continuous medium is reacting by a strain to an imposed, time dependent stress. This field is not passive, and the medium through which it is transmitted isn't either.

This was the most important prediction of Maxwell's unification, the feature that distinguished it from any other theory for electromagnetic phenomena: for fields that vary in time there *must* be electromagnetic waves. The near match of the value for $\epsilon_0\mu_0$ with c^2, the speed of light squared, was strikingly successful in itself. But the more significant *predictions* of the theory were that there must be polarized electromagnetic waves that move at the same speed as light. The experimental verification of this prediction was accomplished by Heinrich Hertz (1857–1894), a student of Helmholtz. Remember that visible effects, sparks, were well known to accompany discharges under the right conditions so there was an obvious connection between light and electrical phenomena. This was not so clear for magnetic phenomena but we'll return to that point in a moment. Since the frequency of the electromagnetic wave depends only on that of the source, which must have a change in time to produce emission, Hertz's idea was to use a simple wire dipole connected to a galvanometer. By discharging a spark on one side of a room he measured the jump in the galvanometer. From the time difference, which is a much harder measurement, he knew the speed of the effect registered by the passive detector. In 1892 he completed the series of experiments that conclusively demonstrated the existence of electromagnetic waves, finding that the speed of propagation closely matched the velocity predicted by Maxwell's theory and more important, the already known speed of light. This was not action-at-a-distance but the *transmission in time* of a disturbance. The technological development of radio was the result of increasing power in the transmitter and improved sensitivity by using resonant detectors, crystals. By the beginning of the twentieth century transatlantic signaling had been achieved by Marconi and radio had arrived.

Radiation Pressure

There was, however, a special feature of this transmission. A boat can be accelerated by a wave when it is hit and the wave scatters. This momentum transfer depends on the energy in the wave and the mass of the boat. If we think instead of a charge being hit by an electromagnetic wave the same thing will happen. Since there is a flux of energy, there is also one of momentum because the propagation is directional. Analogously, a consequence of the electromagnetic unification was the prediction of a new force due to the incidence of light on matter. You might think this was already built into the Newtonian construction for light, particles moving at a speed depending on the medium that carry momentum, but this was never explicitly explored amid the flurry of applications to optical phenomena and was largely overshadowed by the rise of wave theory after the start of the nineteenth century. Once a comprehensive physical/mathematical derivation had been achieved to support the wave picture, it was clear that the flux of energy transported by the waves can impart an impulse. The decisive discovery was by John Henry Poynting (1852–1914), after the publication of the *Treatise on Electricity and Magnetism*. Since in a vacuum the wave propagates at a fixed speed and the vibrations are in the plane perpendicular to the direction of propagation, Poynting was able to show that the *flux* of energy depends on the product of the electric and magnetic field amplitudes and is in a direction perpendicular to the plane of vibration, and the momentum imparted to a particle on which this field is incident is the flux divided by the speed of propagation. Now there was, finally, a direct connection with forces in the Newtonian sense, although not by steady fields. In this case, a time dependent field will move the particle in two senses, around the direction in which the wave is propagating and along it depending on the inertia and charge of the particle. We have now an impulse that depends on the cross section of the particle to the interaction with this wave. In other words, light exerts a pressure.[6] We now know that the Poynting flux does, indeed, accelerate bodies, including satellites and even dust grains orbiting the Sun, an effect first discussed by Poynting himself and later elaborated in the relativistic context by H. P. Robertson almost 50 years later.

Another effect of radiation pressure connects electromagnetic theory to thermodynamics; the link was realized by Boltzmann in the 1880s. Imagine a completely enclosed, opaque container containing only radiation and perfectly absorbing and emitting walls. This is a *blackbody*, an idealization of a body that both absorbs all light incident on its surface and then reradiates it as it heats with a distribution that depends only on temperature until, when equilibrium is reached between the rates of emission and absorption, the radiation (whatever its original spectrum) has precisely the same energy density, ϵ as the walls. Now since this is an isotropic process, the radiation energy density is only a function of temperature, T, and the total energy is simply the energy density times the volume of the box, $E = \epsilon V$. Being adiabatic, the process occurs at constant entropy and therefore the work done by the radiation is precisely balanced by the change in the heat content of the walls. Thus the radiation pressure, which is normal to the walls, is $P_{\rm rad} = \frac{1}{3}\epsilon(T)$. Boltzmann showed, by a purely thermodynamic argument, that $\epsilon(T) = \text{constant} \times T^4$; this agreed with the already known empirical law from Stefan's studies of thermal radiation. It followed, therefore, that the radiation would also exert a pressure on the medium whose ratio to the gas pressure increases as $P_{\rm gas}/P_{\rm rad} = \epsilon_{\rm gas}/\epsilon_{\rm rad} \sim \rho/T^3$ where ρ is the mass density. Returning now to our discussion of stellar structure, this had an additional application—the structure of very luminous stars may be dominated by radiation pressure even when the stars are obviously composed of gas. This last point was exploited by Arthur S. Eddington in the 1920s and 1930s to extend the solar model to the most massive, therefore hottest, stars. So what started out as a unification of two forces of nature becomes a means for understanding the structure of the most massive bodies that emit the radiation.

Acoustics

The development of acoustic theory occurred parallel to electromagnetic theory as a sort of proxy study. An acoustic wave is a pressure disturbance that propagates as a wave in a compressible medium, e.g., air. If the gas is in contact with a vibrating surface, for instance a plate, it responds with successive rarefactions and compressions to the fluctuations in the driving surface. Unlike light, these don't move at a constant speed unless the medium is uniform, and unlike electromagnetism the waves can be driven by a monopole (that is, a pulsating sphere that varies only in its radius), but all of the physics is there. The waves damp because of molecular motion, the Navier-Stokes treatment of the viscosity dissipates energy as heat, but the equations of motion in an ideal medium are identical with those of an electromagnetic wave. It should, therefore, not be surprising that most of the founders of electromagnetic theory, at some time in their scientific lives, studied fluid and/or gas wave phenomena. Lord Rayleigh (John William Strutt, 1842–1919), in particular, was especially concerned with this, founding the modern science of acoustics with his *Theory of Sound* and hundreds of separate studies of topics from wave production to radiation patterns for musical instruments. But it's also about forces and how the boundary conditions change the solutions, describing how different devices, or subsystems, couple together. Sound is a pressure disturbance and therefore the force depends on the area for a given amplitude. Fluctuations in density, or pressure, can be in resonance if their wavelength is a rational multiple of the size or length of a cube. In EM theory, this is the same as a waveguides. If we send a wave down a conducting, or dielectric, channel, there will be resonant modes that depend on the shape of the waveguides. These depend also on the dielectric constant and the magnetic permeability of the walls. Thus, a horn that produces a set of overtone pulsations of air has an analogy with a waveguides in a conical or horn shape for an electromagnetic wave. If we assume, as Maxwell and his contemporaries did, the ether is a continuous elastic medium, it is possible to *analogously* treat the propagation of sound and the electromagnetic field. Even the concepts of wave pressure, of wave momentum transfer, and of harmonics and nonlinear scattering (when the waves interact, strong waves will generate sum and difference frequencies, a ophenomenon discovered by Helmholtz and now the basis of much of nonlinear optics), and many analogous effects find their correspondence in acoustics. In effect, it became a model system for Rayleigh and his contemporaries and later, especially during the effort to develop radar during the World War II (at the Radiation Lab at MIT, the team led by Philip Morse and Julian Schwinger) the analogy became a powerful tool for predicting optimal forms for the transmission lines of the waves and developing resonant amplifiers and detectors.

Dynamics and Electron Theory

Recalling now Faraday's result for induction by dynamo action, we know that it doesn't matter if the source of the field or the receiver of the action is in motion. This is the starting point for H. A. Lorentz's (1853–1928) extension of the principle to the motion of charged particles. A current is deflected by a magnetic field, this is the Hall effect that was discovered by Edwin Hall (1855–1938) and studied extensively by Henry Rowland (1848–1901). Hall found that a magnetic field placed across a conductor produced a potential difference transverse to the electrodes of the flow of the current. It is another manifestation of induction. The variation in resistance identified the charge of the carrier of the current, although this did not indicate whether the current was truly fluid or particles. Even before the detection of charged fundamental particles, the electron, the theory had been fully developed for their motion. He showed that in a magnetic field a charge will experience an acceleration that deflects its motion in a direction perpendicular to both its velocity, v, and the direction of the field, B, writing this as a force,

$$\mathbf{F} = q\left[\mathbf{E} + \frac{1}{c}\mathbf{v} \times \mathbf{B}\right]$$

including an electric field, E, for a charge q. For a deviation from parallel motion, this force produces an orbit around a region of constant magnetic flux, Faraday's field line. This is true whether the field is steady or depends on time and/or position. The coupling will be the charge, q, and the particle will therefore execute a harmonic motion around some direction defined by the field while it moves as well freely along the field if initially so directed. Lorentz then showed that the magnetic and electric fields can be transformed one into the other by a change in reference frame in a way that preserved the wave equation. At the end of the nineteenth century, Lienard and Weichert introduced the time delay when computing the potentials of moving charges, completing the electrodynamic program begun by Helmholtz in 1881. The principle is that we don't see a charge where it is now but where it was some time in the past that depends on its distance and the speed of the signal, the speed of light. Thus there is always a time delay between the cause—the motion of the charge—and its effect—the acceleration of a neighboring charge. This delay produces the emission of an electromagnetic wave and changes respond with a phase shift. This means the motion of the source must be included and leads directly to our next topic, the theory of relativity.

NOTES

1. A complication was that the effect of the Earth's magnetism was always present and Oersted knew no way to shield his detecting needle from its influence but the measurements were sensitive enough to at least show the effect.

2. Julius Robert Mayer had already written a more speculative, although also fundamental, work on the connection between force and work in 1842 but it lacked the mathematical precision and scope of Helmholtz's paper. At the time he composed it, Helmholtz was unaware of Mayer's work. When he later became acquainted with it, in 1852, he acknowledged Mayer's priority in the basic idea of energy conservation. Joule, Thomson, and Tait, among others, disputed this priority for some time, in a manner typical of the nineteenth century, but Tyndall later became the "English champion" of the work and Mayer was awarded the Copley medal of the Royal Society near the end of his life, in the 1870s.

3. The task of setting up standards for electromagnetic measurements and devices occupied much of the late nineteenth century, extending well into the twentieth century (finally, after 1940, a consistent set of international standards were defined but even these changed their values with the shift from cgs-Gaussian to MKS-SI units in the 1960s.

4. In fact, Maxwell had taken years to edit and oversee the publication of Cavendish's scientific papers.

5. As an aside, you should bear in mind the whole of electromagnetic theory developed in Maxwell's hands as a continuous medium. When he speaks of charge he means a fluid, or fluids, that are intermixing and displacing. There is no elementary unit of massive particle, indeed although a confirmed atomist Maxwell was not able to formulate a particle theory of electricity nor did he require one. It's very tempting now to reinterpret this in terms of charges, you know these too well from a century of pedagogical production in the modern curriculum was a treatment . The introduction of a displacement *current* now meant that the magnetic field originates not only from the free movement of the fluid, which can after all be in steady state, but also the specific time dependent reaction of the strained ether. This happened within a medium, the ether for "free space" or a magnetizable substance, the qualitative description of which is that "a change in the strength of the magnetic field, in time, is determined by the displacement current plus the free current."

6. Testing this prediction, which actually is best understood in an astrophysical context produced an amusing sidelight during the late nineteenth century. For some time, a debate raged about whether the effect had been directly observed using a rotating vane the sides of whose blades were painted alternately black and white. The idea was that reflection should produce a greater momentum transfer than absorption because of the impulse felt when the wave direction is reversed coherently. Thus the vanes should rotate in the sense of the light colored surfaces. The opposite was seen, however, and this was finally understood as a thermal effect in the not quite perfect vacuum of the evacuated bulb. But it illustrates how a fundamental idea transforms into a laboratory test that may not be well specified at the start.

8

THE RELATIVITY OF MOTION

> In the last few days I have completed one of the finest papers of my life. When you are older, I will tell you about it.
>
> —Albert Einstein to his son Hans Albert, November 4, 1915

Electromagnetic theory achieved a unification of two forces but at a price. It required a medium, the ether, to support the waves that transmit the force. But this takes time and there is a delay in arrival of a signal from a source when something is changing or moving during this time interval. How do you know something is moving? Remember that Aristotle used the notion of place within space or memory to recognize first change and then motion. So there the motion is known *by comparison* to a previous state. But this isn't the same question that recurs in the seventeenth century. According to Descartes, we have to look instead at the disposition of bodies *relative to which* motion is known. These aren't the same statement, although that would again divert us into metaphysics. Instead, we can look at what this means for the force concept. As Newton was at pains to explain, inertia requires two quantities, velocity and mass, to specify the quantity, *momentum* that is conserved. While mass is both a scalar quantity and a primitive in the dynamical principles, the momentum is not. It requires a direction and therefore, if we identify the inertial state as one moving at constant *velocity*, we need to know relative to what we take this motion. This is where we begin our discussion of the revolutionary developments at the start of the twentieth century: time, not only space and motion, is relative to the observer and dependent on the choice of reference frame and, consequently, has the same status as a coordinate because of its dependence on the motion of the observers. It follows that the notion of force as defined to this point in our discussions must be completely re-examined. Since force is defined through acceleration, which depends on both space and time, without an absolute frame the whole concept of force becomes murky. This is the problem with which Albert Einstein and Hermann Minkowski started and where we will as well.

THE RESTRICTED, OR SPECIAL, THEORY

Galileo, in the *Dialogs*, used a principle of relative motion to refute the arguments of the contra-Copernicans regarding the orbital motion of the Earth, a notion that is now referred to as *Galilean invariance*. If we're in a moving system, and for the moment we'll assume it is inertial in the sense that it moves without accelerating, any measurement we make of forces in that system will be unaffected by its motion. Otherwise stated, there is no way we can find the motion of the system by force measurements on bodies *in* the system. The example he used, dropping a body on the Earth, was based on simple low velocity experiences that are perfectly valid as a first approximation. Thus the location of a body moving in an "enclosure" is seen by a stationary observer as a sum of the two motions. Calling x the position in the stationary frame and x' that in the frame moving with a constant relative speed v, the relation between the two descriptions of the motion in a time interval t is $x' = x - vt$. The time, t, is the same for the two observers, a universal rate of ticking of a cosmic clock that they both agree to use. This could also be two identical pendula; whether on a moving Earth or an observer at rest with respect to the fixed stars, the pendulum will swing identically and therefore the only way to tell which frame the measurement was in is to ask some other question than what is the law of the pendulum. You know from our discussion of the Coriolis effect that there *is* a way to distinguish these systems, the one that is not moving inertially will see the pendulum precess with a period related to the rate of translation of the moving frame relative to the inertial observer and therefore will experience a force. But the time interval is the same for the two observers. The force law is altered by the addition of these *inertial forces* (to use D'Alembert's construction). This was already included in Newton's laws of motion since the equivalence of states of rest or inertial motion guarantees Galilean invariance. Einstein's conception was both more radical and simpler. More radical because he allowed the time to be also something varying for the two observers, and simpler because he asserted in an even stronger form the basis of Galilean invariance: that the physical laws are identically measured by any two inertially moving observers. This second assertion doesn't obviously yield the first as a consequence except for one additional assertion, concordant with experiment. The two observers must agree, somehow, on their relative motion. That they are moving can be asserted relative to another, third, body but that gets us into an endless regression. Instead, we use *as a fact* that the communication takes place by signals at the speed of light and this (finite) speed is independent of the motion of the observer.

It began with a question from Einstein's youth. At age 16, he pictured looking at a clock while moving away from it. Since you see the face and hands because it is reflecting, or emitting, light, if you move slower than the light you will see the hands move (now we would say the numbers change on a digital clock). But if you were to move at the speed of light, if you looked at a watch on your arm you would see the hands moving but you wouldn't when looking at the stationary clock. It would appear to you that time was frozen yet in your moving reference frame you would be aware of the passing of the hours. Now, with the result that the speed of light is actually independent of the observer, Einstein imagined how to extend this with a simple construction. Imagine, instead, that the moving observer is holding a mirror. The image she sees is a reflection of a light beam. His question becomes "does she see her face if she's moving at the speed of light?" If the speed of that light is independent of the motion of the observer plus mirror, the answer must be "yes." But he now asked what someone sitting on the ground watching her whiz by will see. Since the speed of light is the same, there must be a difference in the time it takes for the signal to traverse the two apparently different paths. Let the propagation be perpendicular to the direction of the moving frame's velocity and call that v. The moving observer measures a time interval Δt to go a distance L. The stationary observer sees the same speed but following a longer path, $\sqrt{[(2L)^2 + (v\Delta t)^2)]}$. This must mean they don't see the same time interval and the difference depends on the relative speed of the moving system. This is the concept of time dilation, that the time $\Delta t' = \Delta t/\sqrt{(1 - v^2/c^2)}$. Further, if we ask what the two observers will measure for the length of the path, it *must* be different because the speed, which is a ratio of length to time interval, remains the same. The effect is completely symmetric—as long as the relative velocity is constant neither observer can tell which of them is moving, only the relative speed matters. This is the origin of the concept of *relativity*, the modifier "special" or "restricted" in the usual name for the theory refers to the requirement of inertial motion. Time and space cannot be absolute, the most important change in the mechanical picture of the universe since the introduction of inertial motion.[1]

The Maxwell equations were the framework in which Einstein first discussed the special principle of relativity. When moving through a magnetic field, a charge experiences a force as we've discussed, that depends on its velocity. In the co-moving system, however, this force doesn't exist. If you move with the current the electric field seems to be static. So it would seem the equations of motion change their form depending on the reference frame. If so, do the equations describing electromagnetic phenomena also change? There is nothing special about two systems, it seems, in one of which the charge moves and the observer remains at rest or the other in which the observer moves past a stationary charge. If the two frames are formally, mechanically, equivalent for inertial observers, you would expect that the Maxwell equations for the electromagnetic field should also be transformable to a general system that is independent of the specific reference frame. Thus the title of the first paper on the theory, *On the Electrodynamics of Moving Bodies*, in which he sought to extend the investigations of 1892 by Lorentz on the same topic. The fundamental difference was epistemological. While Lorentz had assumed an ether, Einstein approached the problem as a transformation between observers. It is in this difference in perspective that we see the profound nature of the program that would eventually lead to a new theory of space and time and forces.

There was also an empirical basis for Einstein's assertion of the constancy of the speed of light. The paradoxical observation by Albert A. Michelson, in 1881 and his repetition and refinement of the experiment with E. Morley in 1887, was that there was no trace in the variation of the speed of light for a device in relative motion with respect to the ether. The apparatus was an interferometer consisting of two perpendicular arms that split a beam of light and then, on recombination, produced interference fringes. The phase of the light at recombination depends on the time of travel across the two arms and, so the reasoning went, because of the relative motion there would be a difference in the time of flight along, and perpendicular to, the direction of motion of the beam. This was not detected to a very high accuracy. It was already known, however, that there is a displacement due to the finite speed of light, the aberration of starlight discovered by James Bradley in 1725 and explained as the relative motion of the Earth with respect to a distance light source. If you're watching a particle, this isn't a problem because it's just the vector sum of the components of the velocity. An electromagnetic wave, however, moved in Maxwell's theory through a medium and it is very strange that there was no effect at all. This was explained dynamically by Lorentz, who sought to maintain a Galilean framework in which time is absolute and only space motions are relative, and postulated a physical change in the length of the arm in the direction of the motion. This may seem an *ad hoc* approach but for Lorentz it was a necessary consequence of the electromagnetic strains produced by the ether when a body is in motion. Einstein had, instead, substituted a new central axiom, within the kinematic theory, of the constancy of the speed of light measured by any observers in inertial systems in relative motion. Since the speed of light, c, is a dimensioned quantity, to assert its independence of reference frame implies a fundamental connection between the way two moving observers agree on their definitions of space and time. We can take this as an empirical *fact*. The second axiom, *causality*, is a bit more subtle. For the entire history of physics up to the beginning of the twentieth century this merely implied a semantic relation between *cause* and *effect*. But for Einstein, this was extended to a formal mathematical statement, although this only became clear with Hermann Minkowski's (1907) geometrical representations, that for what is actually a unified *spacetime*, the causal spacetime interval measured by two observers between events a spatial distance dx apart and separated by an interval of time dt is $ds^2 = c^2dt^2 - dx^2$, is constant independent of the motion of the observer. This is a generalization of the geometric concept of distance: the motion is in time as well as space and the total distance covered is in the spacetime. There is a very important difference, however, with the usual Pythagorean theorem, the sign of the constituent terms. Instead of being a simple sum of squares, the space and time components sum with opposite signs, so if you are moving with the speed of light the space interval is essentially the same as the time (with c being a dimensioned proportionality constant). The next step was Einstein's extension of Galilean invariance, for which a position in a stationary reference frame can be related to that in one that is moving at constant velocity by $x' = x - vt$, if the relative velocity v is sufficiently slow. The *direction* of the motion selects which coordinate requires transformation. Since the time is relative, by extension so is the space interval. Einstein then postulated a unique, linear transformation law that would combine space, which I'll denote as the coordinate x and time t and then asserting that the interval measured in the two frames is the same, $(ct)^2 - x^2 = (ct')^2 - (x')^2$, he obtained the same result that Lorentz and Fitzgerald had found from a completely different analysis of the Michelson-Morley results:

$$t' = \gamma(t - vx/c^2)$$
$$x' = \gamma(x - vt),$$

where $\gamma = 1/\sqrt{(1 - v^2/c^2)}$, now called the "Lorentz factor." There are no actual changes in the structure of the body because of the motion but only *apparent* changes due to the causality condition and the constancy of c. The two observers *see* things differently though neither is "correct" in an absolute sense. Einstein added the concept of *simultaneity*, the requirement that two observers *can* agree if they are relatively appropriately situated. By asserting its possibility, he introduced an objective reality into a relativistic construction, the event *does* happen and it's up to the two witnesses to agree on how to coordinate their observations. A striking consequence is the change in the addition law for relative motion. Instead of the Galilean form, he found that if a speed v is seen in a frame moving with a velocity u, instead of the relative motion being $v + V$ it becomes $v' = (v + u)/(1 + uv/c^2)$. In all cases, *this is a direct consequence of keeping the constant speed of light for all observers—no signal can propagate causally faster than c and two observers will always, regardless of their relative frame motion V, see the same value for the speed of light.* It then follows that no body with a finite mass can travel at a speed c.

Let me digress for just a moment. The fundamental picture of the universe containing an absolute space and time relative to which, independent of the observer, all motion can be measured had not changed for two thousand years. It had been elaborated differently through the centuries but had even underpinned Newton's development of mechanics. The idea that all physical experience is only relative and that there is no longer a separation between the motion of an observer and the observation was a rupture in the foundations of physics. Every change in ideas about motion has provoked a change in how force is understood but Einstein had now replaced space and time with a new entity, *spacetime*, so designated by Minkowki. To pass from kinematics to dynamics, however, required another postulate, again related to the speed of light. The two observers must be able to agree on the measurements they're making. That requires some way to establish what it means to do anything simultaneously. At some moment in space and time the two observers must be able to coordinate their measurement device together. Further, their measurements should be independent of which frame they're in. This is the principle of *covariance*, that the physical laws are independent of the state of relative motion of the observers. But this comes with a price. If in one frame a force is applied to an object, in the other it will not seem to be the same. Acceleration resulting from the action of a force depends on the mass of the affected object. But acceleration is the change in a velocity in an interval of time and a force is the change in momentum in the same interval. If the length and time are changing to maintain constant c, then one observer sees a different acceleration than the other and must conclude that the momentum is changing differently. Even more clearly, if a body is accelerated by a known force, then as its speed increases so does the discrepancy between the force applied and the observed acceleration. This implies that the *mass* is increasing since the acceleration decreases. The only thing we can define is the mass at rest, once moving the *inertia* of the body increases. The limiting speed is c and therefore when that is reached the body would not accelerate at all, it's inertia would be effectively infinite. This is what Einstein presented in an amazingly short companion paper to his electrodynamics, "Does the Inertia of a Body Depend on its Energy Content?." He showed that the new dynamics required a redefinition of mass. What we measure is $m = m_0/\sqrt{(1 - v^2/c^2)}$, where m_0 is the rest mass. But defining "rest," remember, is relative so this holds equally to the two frames when the coordinate transformations are applied. He then showed that at slow speeds, the kinetic energy we measure is $\frac{1}{2}mv^2$ but with an additional term, a zero point called the "rest energy" that had been previously unsuspected, and is now the iconographic statement of the equivalence of mass (inertia) and energy, $E = m_0c^2$. Because of the relativity of space and time, energy must also be relative and what you think is the kinetic energy is actually different than what you can extract from actual collisions. If this applies to moving systems, Einstein realized, it also applies to any accelerated motion, in other words even if something is bound in a potential you measure less mass in a static case than you will be able to extract dynamically.

GRAVITATION AND GENERAL RELATIVITY

Inertial motion is very restrictive and Einstein understood that the special theory was only the first step. The struggle to complete this program took over 20 years, culminating with the publication of the field equations for gravitation, General Relativity Theory (GRT), in 1916. The generalization of the relativity principle to accelerated motion provided the first fundamentally new conception of gravitation based only on motion and geometry. At the end of the nineteenth century, Baron Eötvös had demonstrated the equivalence of gravitational and inertial mass. This is the same thing as saying that regardless of the force applied the resistance to acceleration, its mass, is the same as measured by its weight, or that the composition of a body really doesn't matter when you are weighing it. Finally what had asserted in Newtonian gravitation was experimentally verified. It meant, in effect, that the program stretching back to Archimedes, the geometricization of force, might yield an explanation for gravity.

The Equivalence Principle

The equivalence principle was the epistemological core of Einstein's thinking, the insight that changed how force is conceived. The example he gave, imagining yourself in a freely falling elevator (sealed and with no information from outside) was a thought experiment in frames of reference. The trajectory of the elevator is a curve in spacetime because it accelerates. No measurement, however, within this local system can distinguish the absence of gravity from freefall. Within the frame the motions are invariant to changes at constant velocity and there are, obviously, ways of producing forces as well (think of stretching a spring while inside). If you suspend a weight from a spring and accelerate the elevator upward (instead of intentionally imagining your impending death), the spring extends when the elevator starts to move exactly as if the weight, or the acceleration of gravity, had instantaneously changed. Schematically, this is the first step to the covariant formulation of any theory of gravity: for Einstein, at least at first, gravitation was the result of not being able to freely fall and therefore any physical measurements made in one frame should be transformable into any other with just a coordinate change. In other words, with the appropriate description of the trajectory, you can be put on a path through a spacetime in which you don't know about a gravitational field. On the other hand, you *are* actually in freefall, and the gradients in the field produce your trajectory. This isn't something happening in a source for spacetime, there's a mass there—somewhere—that is distributed in such a way that you experience changes in its potential. Think again of the equipotential surfaces. Their normal gradient gives the acceleration but the divergence of this through space is the result of a source. No source, no divergence. Unless, that is, there is a unique point source and we're far enough away from it, or in the proper geometry, that the local effect can be made to vanish merely by displacing our starting point far enough away that the divergence becomes infinitesimally small.

Much of this we already know from the approach that started with Green. The Poisson equation must transform into geometric terms when we enforce this condition of covariance, as must the Laplace equation. If we identify the potentials of the field with the metric on the surface, along with Einstein, then we can say that the curvature (which is measured by the divergence of the normal gradients) is produced by the mass distribution. We can talk—again—in purely geometric terms about gravitation and completely ignore the individual motions thus produced. This may seem very counter-intuitive since you have, up to this point, been thinking in terms of Newtonian gravitation. But one of the main features of general relativity is to render all phenomena as contrast between inertial and accelerated motions and show how to transform between the two.

To extend this idea, imagine now having a ping-pong table placed inside the cab. Again you watch but this time there are two robots that hit a ball repeatedly across the table. You check their activities to be sure the ball is perfectly elastic and they're always hitting with the same stroke. Then at some moment, you see the ball passing between them not in an arc but directly across the table. By the reasoning of the equivalence principle, we know what has happened: the elevator is in freefall. In the local frame there is no net force and the motion appears purely inertial. So even if we have a non-inertial motion, we can transfer into a coordinate system that permits solution of the dynamics as if there were no forces, and then find how the frame is accelerating and reconstruct the trajectories in *our* reference frame.

Gravitation as Geometry

First let's begin with a simple example. You want to make a map using a group of surveyors each of whom is dispatched to a different city. To be specific, take New York, Baltimore, Chicago, and Boston. On arriving at their stations, each sets up a local coordinate system, establishing the vertical as the direction perpendicular to the horizontal (which may seem obvious but wait a moment), and measures the angles between her city and two others, for instance the angle measured at NY between Baltimre and Chicago. After they have closed this net of cities, these observers compare the sums of the angles. For a tri-city set, a triangle, they know from Euclid that the sum should be 180 degrees. But when they do the sum, it isn't. What are they to conclude about the surface? This problem, posed precisely this way, occupied Gauss at the start of the nineteenth century and produced a first picture of surface curvature: in the measurement is made purely geometrically,

Figure 8.1: Albert Einstein. Courtesy Library of Congress, Prints and Photographs Division, LC-USZ62-60242.

and the result for all possible triangles differs from the expected sum by the same amount, the surface is *curved* and that curvature is uniform. The radius of curvature, the degree to which the measurements depart from the Euclidean flat plane, is found by the dependence of the difference in the sum on the metric separation of the points of reference in the global system. Any rotation of the triangles, any shuffling of the measurements, will produce the same answer if the surface always has the same topology, the same "form" of curvature. As a further point, if two observers establish a baseline and then translate perpendicular to that, always carefully establishing a baseline after some parallel displacement without checking the distance, they will either relatively converge or diverse. If they ultimately converge, the surface is positively curved and the simplest example we know is a sphere. If they diverge, the curvature is negative and, geometrically, we can imagine it is saddle-shape.

Now we ask a different question, how does each surveyor establish the vertical. They might simply put a piece of paper on the ground and find the line that makes an angle of 90 degrees to it and doesn't lie in the plane. But more likely, they will use a plumbline. Now they might notice that these two don't match and we've returned to the concept of the potential as a surface. The normal component of the potential gradient, which defines the local gravitational acceleration, depends on the local distribution of masses. This was the result of the Maskelyne measurement,

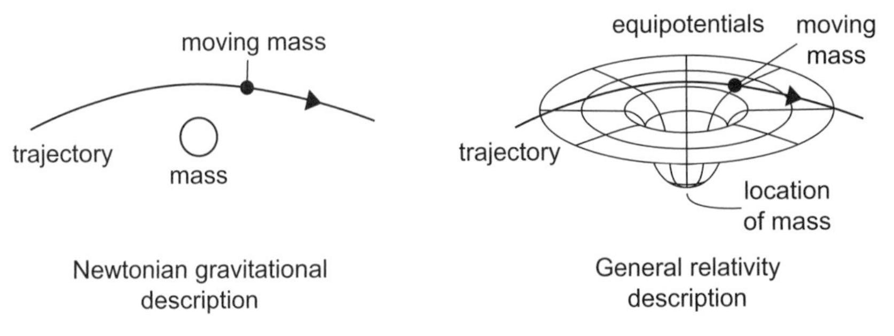

Figure 8.2: A cartoon comparison of the classical (Newtonian) and relativistic descriptions of the deflection of a body by another mass. In the left panel, the trajectory is changed because of a force. In the right, one can describe the motion along a curved spacetime (here shown schematically as a potential well but the motion is really in four dimensions, the time changes along the trajectory).

the deviation of a plumbline because of the vicinity of perturbing masses (i.e., mountains in his measurement). But if the surface is apparently level, in the sense that there are no local variations in altitude, we would conclude that the vertical deviates because the Earth's substructure is not of the same symmetry as we assumed for the curvature based on the geometry. So we have a way to see how curvature defined by a measurement of the potential gradient relates to that defined by the local tangent plane, the surface geometry.

Let's go one step farther. If we again take two observers, one of whom stays fixed while the other walks in a straight line, that is constantly noting that the stationary observer always stays at the same bearing relative to an instantaneously established coordinate frame (again, measuring the local vertical and the angle relative to the original bearing of the trip), the two may notice that the two vertical directions are not translating parallel to each other. That is, the normal vector may rotate in the plane perpendicular to the surface. If the two stay aligned, the surface is obviously fat in the euclidean sense and we are simply moving in a uniform straight line. But if the angle changes, increasing in a positive or negative sense, the surface is curved. Again, we can measure this using a plumbline but that introduces a new feature, the local gravitational field may not be following the surface as we define it by the path. The parallel translating observer thinks she is walking a straight line. In other words, if she maintains constant speed and direction he is moving *inertially*. But the stationary observer sees that this isn't so, he concludes there has been an acceleration even if the movement is at constant speed (for instance, the two can exchange a signal at fixed frequency to check that there has been no change, hence no acceleration measured by the Doppler effect) since the two normals are rotating. Further, if the moving observer then translates around a closed circuit always maintaining the same heading, without rotating the coordinate system, then at the end of the circuit when the two are reunited there will be a rotation of the coordinate patch with which the moving observer recorded the original directions when starting the trek. In moving around a closed curve—that is, an orbit—around a central mass in the two body problem, the orbiting body is moving at constant speed and parallel translating. But this is clearly an orbit because if we translate the velocity vector back to the starting point it will have rotated relative to the radial direction by an amount that depends on the angle through which it's been displaced around the orbit. The change, Newton realized, will always be in the radial direction and, since by the second law any acceleration is in the same direction—and with a magnitude that depends on—the force producing it, we interpret the change in the velocity vector as a result of a force.

This is the equivalence principle in another guise. Any acceleration is interpreted—based on this geometric observation—as a force and we have now completed the analogy that Einstein and Marcel Grossmann (1878–1936) achieved in their 1913 paper on gravitation, *Entwurf einer verallgemeinerten Relativitätstheorie und einer Theorie der Gravitation* ("Outline of a General Theory of Relativity and of a Theory of Gravitation"). Because this is a local effect, constructed by taking pieces of the path and moving them, they had constructed a differential geometric description of the local curvature that is equivalent to a force. The deviation of the normal gradient to the equipotential surface is the same as a differential force, this provided a description of gravitation: a gravitational field can be represented using differential geometry as a curvature. Since this happens in both space and time because of the relativity of a spacetime, this is a curvature in four dimensions, the basis of the General Theory of Relativity for gravitational fields.

The direct influence through the Einstein–Grossmann collaboration was Bernard Riemann's 1857 thesis *On the Hypothesis at the Foundation of Geometry*, a topic chosen by Gauss at Göttingen. In this work, Riemann introduced the idea that a differential expression for distance can be generalized for a non-Euclidean surface or space, maintaining the same form but allowing the *metrical* distance to depend on position. In Grassmann's or Hamilton's language, this is the same as saying the scalar product—that is, the projection of two vectors onto each other—depends on the position in the space. The projector is dimensionless if the two vectors are unit distances and, instead of the Pythagorean theorem, that square of the total of small displacements is the sum of the squares of the component displacements (hence using dx for each component),

$$ds^2 = d\mathbf{x} \cdot d\mathbf{x}.$$

Riemann generalized this using a projection factor, the *metric*, which is a matrix that also includes the displacements in the plans and not only along the individual axes,

$$ds^2 = \sum_{i,j} g_{ij} dx^i dx^j,$$

where we sum over all possible values of the two indices i and j (for instance $g_{11}(dx^1)^2 + g_{12}dx^1dx^2 + \cdots$). In words, this says that for two infinitesimal displacements, instead of being the sum of the squares, the distance may change depending on how the two distances displace parallel to each other. The change in the properties of the surface can then be determined by examining how the metric, this projector, changes as we move around the space. The relation between this *metric* and the surface (or spatial) curvature was made more precise by Gregorio Ricci-Curbastro (1853–1925) and his student, Tullio Levi-Civitta (1873–1941), during the first decade of the twentieth century. Specifically, Levi-Civitta was thoroughly familiar with dynamics, especially rigid body motions, and introduced the idea of parallel displacement to describe motion on generalized curves. Luigi Bianchi (1856–1928) further developed this "absolute calculus" by finding the relation between the scalar quantity—the curvature of the surface, and the local measurement of the deviation of the vectors by this transport.

Grossmann, who had been a fellow student with Einstein at Zürich and was now a professor of mathematics at the university there, knew this part of the mathematics and Einstein had the intuition for how to apply it. Their principal joint paper, the *Entwurf* paper of 1913, actually consisted of two separately authored parts, one on the methodology (Grossmann), the other on the physical setting (Einstein). While still incomplete, this was the crucial start to what proved to be a difficult, four year endeavor for Einstein alone. Let's return to our surveyors. Recall we used the

Doppler shift to establish that the speed is constant between the two observers. It doesn't matter, really, if one is truly stationary or not, they are simply in relative motion as in the Special (restricted) Theory of Relativity. Now in sending the signal, if the rotation of the normal vector is due to a gravitational field, any acceleration must produce a change in the length *and* time intervals. If the speed of light remains constant, the alteration of the normal gradient to the equipotentials will produce a shift in the frequency of the signal, if it emerges from a stronger gradient, if the normal direction changes more rapidly, the signal will be received with a redshift—a downshifting of the frequency. It also means, because the path of the light is constrained to maintain constant speed, there must be a curvature to the path even in a vacuum irrespective of the nature of the light and even if the pulse has no mass. In Newton's conception of light, a particle with a finite mass, a gravitational field produces a deflection when it isn't along an equipotential. But it's a different effect. If light is photons, and these are massless, their constant speed requires again a deviation but with a value a factor of two different from the Newtonian prediction. Einstein realized this even before the theory was in its final form, as for the redshift, and suggested after discussions with an astronomer at Göttingen, H. Freundlich, that this might be observable as a deviation of the positions of stars near the limb of the Sun during a total solar eclipse. Now we can ask what is the source for this change in the trajectory? In the two body problem we identify it with the mass of the two orbiting bodies. The closer they are together the greater the gravitational acceleration, therefore from our new perspective the greater the curvature of the spacetime. Since we already know from the restricted theory that the inertia and energy are identical, the source can be generalized to the total energy density.

You can now see why Einstein followed this path. In a gravitational field all bodies experience the same acceleration regardless of their mass. The difference is the force. They follow identical trajectories, that is freefall, and we can always find a frame of reference—the moving observer—for whom the principal gravitational field of a central body vanishes. But there is also an effect of the tidal component, the higher order derivatives of the gravitational field, that never vanish regardless how small we make the region. Thus for a strictly local measurement it is reasonable—the elevator *gedanken experiement*—to imagine an observer in freefall feeling locally no gravitational acceleration in whose reference system there is only inertial motion (or effects produced strictly by interactions within that frame). This is motion along the normal gradient to the equipotentials. Along a path we can define an ordered sequence, a single monotonic parameter that describes the steps. Since the observer receives and sends signals to others, the reception times depend on the spacetime as a whole but she can look at her watch and note the times. This is the *proper* or *comoving* frame, the time in this frame is therefore an adequate *affine parameter*, the sequence. The observer defines any motion as *covariant* relative to the *contravariant* background.

We return to Green's conception of the field. A fluid mass adopts a shape that corresponds, absent internal circulations, to the equipotential surface. This may be substantially distorted from a sphere depending on whether the mass is rotating or isolated. In a close binary system, for instance, the Roche solution to the potential produces a cusp in the shape when the local gravitational forces balance. This can be easily seen in the translation of a mass over the surface. On any equipotential the motion will remain constant so this is inertial, but in a non-inertial reference system such as a rotating body, the normal cannot remain constant on the equipotentials. In a more complex mass distribution, the local curvature thus reflects the distribution of the masses and we have finally finished. Instead, in general relativity, the geometry *is* the field. This is why Einstein persistently referred to the metric as the field itself, with the curvature taking the role of the field equations. The field equation must reduce to the Newtonian limit, therefore to the Poisson equation, when the fields are sufficiently weak.

Orbital Precession and the Two Body Problem

Planets orbiting the Sun follow closed trajectories, at least in the original form of what we've called the Kepler, or two body, problem. The period depends only on the angular momentum of the body and its distance from the central mass. Now, remaining within the frame of special relativity, we think about the velocity of the planet. When it is near the Sun, for example at perihelion, it's moving faster than at aphelion. Comparing two instants, the inertial masses will not be identical and the angular momentum can't be the same either. Although it's stretching the limits of special relativity to describe it this way, to a truly inertial, stationary observer watching this the planet will appear to move as if affected by some other force, or by something other than in inverse square central attraction. The orbit, in fact, cannot be strictly periodic and must precess, exactly as though some other body were perturbing its orbit. The change, due to the relativistic effect, depends on only one quantity, the orbital speed. Since this is always small compared to c, we can be sure the effect will not be very large. But neither does it absolutely vanish and thus, after all the planetary perturbations in a real solar system are accounted for, there should be a residual apparent force that somehow is related only to the central body. Throughout the nineteenth century, progressively more precise measurements of the motion of the planets were achieved. Because the era of telescopic measurement had, by the end of the century, spanned more than 150 years, for the shortest period planet—Mercury—it was clear that a cumulative apsidal motion that could not be simply explained by either measurement errors or the perturbations of the planets was present as a residual in the data. Small though it was, about 43 arcseconds per century, it was nonetheless far in excess of the typical measurement errors of a few arcseconds, and it was increasing secularly. It was natural to try the same explanation that had proven so spectacularly successful in the discovery of Neptune, that another body was present within the orbit of Mercury whose gravitational perturbation was strong enough to affect Mercury without altering the orbits of the neighboring planets. Not surprisingly, it was Leverrier who attempted this and predicted the orbit based on the Mercury anomaly of a body, subsequently named "Vulcan," of low mass and very short period. That the body was sighted several times during the last decades of the century is not important, it doesn't exist and that alone eventually ruled out this obvious explanation. One other possibility remained, the same one invoked by Newton to explain the lunar apsidal motion: the Sun might not be spherical. As we have seen, the gravitational potential of a sphere is different in its radial dependence than a spheroid. Regardless of the origin of this presumed eccentricity of the Sun, were it homogeneous and differed from a sphere by only a few arcseconds (about 10^{-5} in relative radial distortion) this would produce an alteration of the gravitational field at Mercury sufficiently large to account for the excess. Again, the observations firmly ruled out any deviation of the proposed magnitude. Lacking these explanations, Simon Newcomb—then director of the U.S. Naval Observatory and the leading celestial mechanician of end of the nineteenth century—proposed a drastic solution, that at short distances the force law showed a small departure from $1/r$ form of the potential, of order one part in ten thousand in the exponent. This has the same effect as a nonspherical Sun but a totally different origin that, he admitted, was purely a hypothesis. By itself, the suggestion shows that the concept of gravitation was still not quite consolidated since all other successes in celestial mechanics should have served as sufficiently strong tests of the force law to rule out this speculation *a priori*.

The explanation is, instead, a product of the relativistic generalization of gravitation. To state this in the language of GRT, in the vicinity of a massive body—for instance the Sun—the curvature of spacetime is changed from that of a classical GM/r potential field to one only slightly different. Since the space is no longer flat, an orbit doesn't close even around a point mass and, as a result, an elliptical orbit slowly precesses. Recall the orbiting particle is in freefall not only motion though space in time, but motion in spacetime, and therefore both the time and space intervals it experiences are different from those of an observer at rest on the central mass. Einstein realized this orbital precession as the first consequence of the weak field limit in the final version of GRT using the calculation from a completely different context. In 1920, Arnold Sommerfeld applied the relativistic correction to the Bohr atomic model in which an electron orbits the nucleus at nearly the speed of light. For such a particle, although its rest mass is constant its momentum is not and therefore neither is its orbital angular momentum. Its inertia increases when "perinuclear" (near the force center) and decreases when "apo-nuclear." The variation produces a precession even for the two body Kepler problem.[2] The change in the field is enough to precisely produce the already known anomaly, a "retrodiction" that proved decisive in Einstein's labors.

The Schwarzschild Solution and Singularities

The field equations were presented in their final form in Einstein's paper of 1916. Their solution for a point mass was obtained by Karl Schwarschild (1873–1916) in the same year, almost immediately after their publication. It was a tour-de-force calculation made possible by a reasonable simplification: for a spherically symmetric point mass the problem is one dimensional. The time (g_{00}) and space (g_{11}) components of the metric depend on the radial position, which is obviously true using the picture of equipotential surfaces precisely because we have a spherically symmetric system. The same is true for the curvature. Because the mass is pointlike its external field is a vacuum so the classical limit for the field equations is the Laplace equation. Schwarzschild then was able to solve the linearized problem and found:

$$g_{00} = \left(1 - \frac{2GM}{c^2 r}\right),$$

$$g_{11} = 1 \Big/ \left(1 - \frac{2GM}{c^2 r}\right).$$

The angular coordinates define motion on a sphere of constant radius. This is because the only change that matters is crossing from one equipotential surface to another, not motion confined to such surfaces; circular orbits do not spontaneously become elliptical. For Mercury, at a distance of about 1/3 the distance of the Earth from the Sun (about 100 solar radii), this amounts to a deviation in both the temporal and radial components of only about 5×10^{-6} from a perfectly flat metric. With an orbital period of about one quarter year, this amounts to a change of about the right order of magnitude for the nonclosure of the orbit per cycle. Einstein had already obtained this result before arriving at the final form of the field equations so this agreement was important. There are two additional consequences. The change in the time component implies a redshift when light emerges from the body and heads toward the great beyond, the *redshift* (infalling light will, instead, experience a blueshift but that is not visible by a distant observer). A third consequence is the bending of light trajectories at any distance from the central mass: since the Sun has a finite size, its limb defines the point of closest approach for an "orbit of light."

Schwarzschild's solution, however, introduced a puzzle. Green had initially insisted that the distribution of charge on a conducting surface remain non-singular, that it be everywhere smooth (and therefore continuous so the derivatives also exist everywhere and there are no infinite forces). Now we have a strange consequence of the relativistic equations: at a distance $R_\star = 2GM/c^2$ from a point mass, the spatial metric component becomes infinite and the time component vanishes. In other words, viewed from afar, a freely infalling observer will simply appear to hang at some distance from the central point mass and disappear in as a progressively more redshifted image. At first thought to be merely an effect of a poor choice of coordinate representation, it is now understood that the Schwarzschild metric possesses an *event* horizon from inside of which it is impossible to receive a signal although the infalling particle experiences nothing unusual (except for the tidal acceleration resulting from the locally enormous curvature). The curvature doesn't vanish even though the metric becomes singular, hence the spacetime can be continued smoothly into the center, although from the publication of Schwarzschild's paper to the early 1960s this was thought to be a mathematical artifact and there were several attempts to remove it by different coordinate representations. The continuity of the spacetime was finally demonstrated by Krushkal and Szekeres in 1965, long after the original derivation of the metric, and the metric for a rotating isolated mass was obtained by Roy Kerr in 1963. With these two solutions, the importance of the singularity and the event horizon were finally recognized and a new name was coined by John Wheeler, a "black hole."

Gravitational Lensing as the Map of the Potential

The most celebrated confirmation of GRT came in 1919, the detection of the gravitational deflection of starlight at the limb of the Sun observed during a solar eclipse. But that is only a part of the story. Newtonian mechanics also makes such a prediction, the trajectory of a body moving at the speed of light (which is not a speed limit in classical mechanics) should be about 0.9 arcseconds at the solar limb (one solar radius for a one solar mass body). Relativity, instead, makes two fundamental modifications, both qualitative, that predict to a value twice the Newtonian deflection. Since time is relative, the frequency of the light is changed on the two legs of the path, first being blueshifted on ingress into the gravitational potential and then redshifted on egress. The net change is, therefore, zero but this changes the effective inertia of the light. The second is the change in the momentum as a function of position in the spacetime, depending on the energy of the light since $p = E/c$. This is a direct measurement of the structure of the spacetime.[3] This optical consequence of general relativity is more, however, than just a test of the theory. It provides a means to image the potential, to visualize the correspondence between the force and the geometry. Unlike a normal lensing device, the vacuum has a constant, uniform "index of refraction" and therefore any change in the trajectory of light is only because of the change in the spacetime *structure*, not its material properties. Recall that Maxwell had assumed the ether has a constant dielectric constant and magnetic permeability. In the transition between two media, or if either of these physical parameters varies within a region, the path of light changes because of dispersion. A wave *in vacuo* doesn't disperse but follows a path of least action, a geodesic, and therefore traces the curvature. Now if instead of a point source we view an extended object through a distribution of masses, the lenser is more complex. Two or more paths may converge at the observer. This has been observed in a setting Einstein did not imagine, but Fritz Zwicky did in the 1930s. The distance between galaxies is huge, of order hundreds of millions to billions of light years, but the sizes of clusters of galaxies are not that much smaller, about one to ten percent. Thus, a distant object whose light crosses one of these clusters is, in effect, still in the near field of the lenser and a focused image can be produced by convergence of rays from more than one direction. The resulting image is, to be sure, very distorted but completely characteristic: since there is no net redshift because although the light must emerge from the vicinity of the cluster, it also entered from a distance, the spectrum of the light is unchanged while its path deviates. The images should have exactly the same spectrum, a wildly improbable vent if one is really seeing merely a chance superposition of two distinct cosmic emitters. Such lenses on large scale have been observed, now almost routinely, since the discovery of the first lensed quasar, 1957+256 (the "name" is the celestial position) in 1979. Lensing by individual stars—microlensing, the closest thing to the original test proposed for GRT—has been possible since the mid-1990s but this is only because of the motion of stars in front of a more distant background and requires a different observing technique. Here the motion of the lenser in front of a densely populated background of stars, for instance toward the bulge in the center of the Milky Way, produces a ring that crosses the Earth's orbit. The brightness of a background star changes as it is lensed in a symmetric way over a timescale of several days. The technique has even detected close, unresolved planetary companions of the lenser since any mass, no matter how small, will produce a deflection and leads to multiple peaks in the brightness variations of the background star.

Gravitational Waves: A New Wrinkle

Classical gravitation has no time dependence. If bodies move the potential changes with their re-ordered distribution. But in GRT, there is a new feature that is again due to the finite speed of any signal. A change in the symmetry of a body, or of the distribution of mass in an ensemble, alters the structure of the gravitational field *in time*, not only in space. You can think of this as a change in the curvature of the spacetime, which propagates away from the source and carries energy (and, momentum). The emissivity of the configuration (we'll say system but this could also mean a single body whose shape changes) depends on the rate at which the distribution changes. Eddington, in the 1930s, realized that this is a necessary consequence of the new theory of gravity but it is *extremely* weak (made accessible only by the possibility that enormous, of the order of stellar, masses maybe involved.

The Hertz experiment, primitive as it was and as early in the development of electromagnetic theory, succeeded because the coupling between electromagnetic waves and matter is so strong. For gravity, in contrast, the coupling is nearly forty orders of magnitude smaller and, it would seem, the detection of such waves as a test of the theory is a hopeless task. In the 1950s, however, Joseph Weber suggested a way to do this using massive cylindrical bars that would resonate due to mechanical strain produced by a passing gravitational wave. Like an EM wave, a gravitational wave is polarized in the place perpendicular to its propagation direction. Unlike a vacuum electromagnetic field, however, the oscillation is quadrupolar and has a distinctive signature of strains in two orthogonal directions simultaneously with the two senses of polarization being at 45 degrees to each other at the frequency Ω. In addition, because of the extremely strong frequency dependence of the generation mechanism, the rate of energy emission varies as Ω^6, only the highest frequencies need be sought and the bars can be made manageable sizes. Several cryogenically cooled bars have been operating since the 1960s in laboratories around the world. Since there isn't a single, fixed frequency expected from cosmic sources the narrow bandwidth of the bar detectors is a serious limitation. A different mode for detection was proposed in the 1980s that uses a Michelson interferometer, much like that used to study the variation of the speed of light due to the motion of the Earth. A laser beam is split into two and sent through a pair of perpendicular evacuated tubes. The beams are reflected from mirrors mounted on suspended masses at the ends of the tubes and recombined to interfere. Actually, the beam reflects a large number of times from both ends of each cavity before recombination to extend the pathlength, thereby increasing the sensitivity because when a gravitational wave crosses the instrument the lengths of the two arms change. The differential change in the path is detected by looking for changes in the intensity of the interference pattern using photoelectric cells. The fluctuations in time of the light signal contains the frequency information with which to construct a broadband spectrum, the current limits being a few kilohertz. Several such instruments, actually full scale observatories, are now operational, the principal ones being a four kilometer long the Laser Interferometric Gravitational Wave Observatory, LIGO, in North America and Virgo, with three kilometer arms, in Italy. Others are located in Australia, Japan, and Germany, and there are plans for constructing such detectors in space on scales of millions of kilometers. Combined with the bars these form a worldwide network. But unlike electromagnetic waves, which required an experimental verification, the demonstration of gravitational wave emission was accomplished in the early 1970s when R. Hulse and J. Taylor discovered the compact neutron star (pulsar) binary system PSR 1915+21. This is a short period system that displays a systematic decrease in its orbital period despite the complete separation of the two orbiting bodies. The decrease in the binding energy can be explained by the radiation of gravitational waves that causes the system to contract.

Cosmology

Having completed the general theory in 1916, the following year Einstein followed the same expositional path as Newton and extended the new conception of gravitation to the biggest problem of all, the structure of the universe. Around a point source, as Laplace had developed the Newtonian picture of the gravitational force, there is no tide for an orbiting point. An orbit, whatever its angular momentum, is always closed. But as we have seen, for a point in the relativistic case there is a variation in the mass (inertia) of the orbiting body and therefore the angular momentum is not constant although it *is* conserved. Thus, as is the case for an extended mass, the central body creates a tide that even a point mass in motion feels. At the cosmological scale, the overall mass distribution, creates a local force that also contains the equivalent to a tidal force according to the Poisson equation. The divergence of the force, for a general distribution, causes this. The same must be true for a relativistic framework since the covariance principle allows us to transform from one reference frame to another. The curvature of an equipotential surface is another way of saying we have a tidal term. For a point source this is clear, as you get close to the central body any finite region of space experiences a second derivative of the potential, the curvature is the inverse square of the length over which this gradient occurs and this is proportional to the density. Thus the timescale for freefall becomes comparable to the light crossing time and at that point we have an event horizon. Einstein's intuition was that the tidal acceleration is the same as a curvature and if the geometry of spacetime is global the local curvature produced by the large scale distribution of masses will yield the gravitational field of the universe and determine the motion of all constituent masses. On an equipotential surface the motion is inertial. But if we have a tidal force, there is a change in the gradient to this surface from one point to another, the surface behaves in space as though it is curved. The important difference between classical and relativistic gravitation is that the spacetime, not simply space, changes. That means there is a possibility that two observers can see a differential redshift of light, not from relative motion but from differences in the local density around each, and the spacetime itself will seem to be expanding. This is the basis of cosmology, the origin of the observed relation between distance and redshift. To go into more detail, you should consult Craig Fraser's book in this series.

IS THIS ALL THERE IS?

Usually in histories, we ignore anomalies and dead ends but some have been so prominent, and so comparatively long-lived that I don't think it appropriate to simply pass over them silently. This brings us to the edge of precision measurement and also a fundamental problem in modern cosmology. While it is beyond the scope of this book to review modern relativistic cosmology (again, I recommend the book by Craig Fraser in this series for more discussion), I will mention the fundamental problem and its solution in a semi-classical framework to again demonstrate the point that both methods and questions persist in physics for far longer than they sometimes seem to.

As we saw, Newcomb proposed a modified gravitational force law as a solution to the perihelion precession anomaly for Mercury. In the 1970s, the rotational motion of disk galaxies was shown to require a significant component of nonluminous matter, now called *Dark Matter*, homogeneously distributed through the system and extending to large distances from the center. In time, this has grown to about 97% of the total gravitational mass. While the nature of this matter is heatedly debated, and likely will be unresolved even as you are reading this, this paradoxical situation has spawned two proposals that bear striking similarities to those predating General Relativity. The first proposal comes from cosmology and is directly linked to the explanation for the structure of the universe provided by general relativity. You'll recall, I hope, the discussion of the virial theorem, which states that the there is a fixed sum for the total kinetic and gravitational potential energies for a bound system. If we imagine a cluster of galaxies instead of a cloud of gas particles, bound to each other by mutual gravitational attraction, classical Newtonian gravitation is sufficient to determine the total amount of binding mass. You measure the velocity of each galaxy and find the sum of the kinetic energies. Then, measuring the projected size of the ensemble, the radius of the cluster, you can compute how much mass is required to keep the system bound. This can be directly compared with the masses of the individual galaxies obtained from their starlight and from observing the motion of gas bound within them. How this is done would take us too far from the point. What has been found after several decades of study, and this will certainly be ongoing even as you read this section, is that the mass required to keep the clusters together exceeds that of the visible galaxies by up to ten times. Because this matter exerts only a gravitational force, hence is massive, but emits no light at any wavelength it is called "dark matter." Its discovery has been one of the strangest consequences of relativistic modifications to gravitational theory because it is not only possible to include it in the description of the expansion of the universe, it is actually required. How this affects our notions of forces depends on what you think it is. There is also evidence that the universe is now expanding at an accelerating rate and this requires a repulsion in the spacetime now. No known force is capable of producing this. Gravitation, as you know, is only attractive and this effect seems to be important only on very large scales, of the order of the size of the visible universe or about ten billion light years, and not in local scale of the solar system, about ten light hours. If this is correct, it points to the most significant modification in our ideas of matter in the last three hundred years.

NOTES

1. One way o see why this is necessarily a consequence of the invariant speed of light is o remember that a speed combines the dimensions of space and time. If the time interval differs between the two observers so must the length interval to compensate. But this also means that as the speed approaches c the finite time interval of the comoving observer becomes an infinitely long one for the stationary observer and vice versa.

2. Sommerfeld used this to explain the fine structure of spectral lines, assuming strict quantization of both the orbital angular momentum and this perturbation. In GRT this is not necessary since the orbits form a continuum but the relativistic mass effect remains.

3. To be more precisely of the metric. Light follows a null geodesic, meaning that in the language of our previous discussion $ds^2 = 0$ so that the components are related by $g_{00}dt^2 = g_{11}dr^2 + r^2d\omega^2$ where $d\omega$ includes all the angular terms for a spherically symmetric spacetime where the g coefficients depend only on the radial distance from the center of the mass. The angular deflection is a consequence of the potential at r and is nearly the same as the ratio of the distance of closest approach, b to the Schwarzschild radius, $\Delta\theta = 4GM/bc^2 = 2R_\star/b$. For a solar mass, $R_\star$ is about 3 km so this amounts to about 1.8 arcsec at the solar limb, a very difficult measurement during a solar eclipse but nonetheless tractable even with the photographic emulsions and telescopes available at the start of the twentieth century. The observational test, finding a value that within the uncertainties disagreed with classical theory and corresponded to Einstein's prediction, catapulted Einstein to public fame as a virtual cult figure. Subsequent solar system tests included the direct detection of time dilation for signals propagating near the limb from satellites.

9

QUANTUM MECHANICS

> There is no doubt that the formalism of quantum mechanics and its statistical interpretation are extremely successful in ordering and predicting physical experiences. But can our desire of understanding, our wish to explain things, be satisfied with a theory which is frankly and shamelessly statistical and indeterministic? Can we be content with accepting chance, not cause, as the supreme law of the physical world?
>
> —Max Born, *Natural Philosophy of Cause and Chance* (1949)

Atomic structure at the end of the nineteenth century was a mix of combinational rules derived from experimental chemistry and simple physical models. The investigations of the eighteenth and nineteenth centuries had been attempts toward the microscale applications of force laws, combining mechanical and electrostatic ideas but more directed toward the interactions between particles than their structure. These two different streams converged during the second half of the nineteenth century mainly through spectroscopic discoveries: the identification of specific unique sequences of emission lines that fingerprinted the elements.

THE ATOM AND THE ELECTRON

At mid-century, Kelvin and Helmholtz had demonstrated the persistence of vortices in a fluid in a celebrated theory dealing with their circulation. This quantity is the integral of the velocity around the circumference of a vortex ring or tube and Kelvin's theorem states that it is a conserved quantity in the absence of viscosity an a constant in an ideal fluid. It means that two vortex rings can pass through each other, collide and bounce, and generally behave as though they are fundamental "elements" of a fluid. In the dynamical theory that resulted from this theorem, expounded most completely by Kelvin in his *Baltimore Lectures* (1884), counterrotating vortices attract through a force of the form $\mathbf{v} \times \omega$ where the *vorticity*, ω, is defined as the circulation of the velocity field $\mathbf{v}$, This is the Magnus force. A kitchen experience illustrates the basic idea. Imagine using an egg beater to whip cream. The rotors turn in opposite directions, accelerating the fluid between them. The flow is described, then, by Bernoulli's law—the pressure drops and the beaters attract. In a world filled with an incompressible ether, it's easy to see how it is just a short step out of the kitchen and into the atomic realm. The stability of vortex rings was especially fascinating, so much so that Kelvin's collaborator, Peter Guthrie Tait, regularly used a smoke-ring generator in his demonstrations. The rings are more intuitive objects than spherical vortices, having already been considered in Cartesian mechanics. But Descartes did not possess any mechanism that could render such structures dynamically stable. Kelvin and others added the possibility for the vortices to vibrate in nonradial (that is, circumferential) modes that naturally produced harmonic series. Thus, if they were electrical bodies, these pulsating vortices could produce electromagnetic emission at specific, unique frequencies—emission lines—although the mechanism is left unspecified. The research program occupied a number of Cambridge physicists, among whom was John Joseph Thomson (1856–1940) for whom the dynamics of vortex rings was the basis of his 1884 Adams Prize essay. By the end of the century, albeit with few adherents, this model appeared viable as an explanation of atomic phenomena.

In contrast, the electrical program led to an atomic model that consisted of positive and negative charges homogeneously distributed and the mass coming from self-interactions. Such a model, whose charging properties could be understood by increases of deficits of one or the opposite charge, was especially successful in explaining the chemical laws. It was an electrostatic, not a dynamical, explanation, an essentially structureless spherical mix of charges that insured neutrality. Its chief explanatory importance was the structure of materials, not spectra, which it could not handle. But it could explain how the electrical properties of gases change when subjected to a sufficiently strong field. When a gas is ionized, it conducts. This change was explained by the liberation of free charge by the exciting agent, whether an electrostatic field, light, or a current passing through the gas. The program of extending forces to the microstructure of matter was completed with the discovery of the electron and understandings its role in the structure of the atom. The three fundamental demonstrations of the existence of this particle still required classical fields. Two of these were similar to mechanical experiments. The third was more indirect, based on spectroscopy.

The first discovery was directly linked to the structure of the atom. Peter Zeeman (1865–1943) found that atomic spectral emission lines split into definite patterns when a magnetic field is applied. He was inspired by a casual remark of Faraday's about the effects of the field on the structure of matter. Specifically, Zeeman found two patterns. Along the direction of the field, there were always two of opposite senses of circular polarization. Transversely, a third appeared between the others and the lines were linearly polarized. This was immediately explained within Lorentz's electron theory by the application of two forces, electrical and magnetic, through the inductive force. As long as the fields are static, the trajectory is a helical path around the direction of the magnetic field line. Acceleration along the field is only due to an electric field. Now since light, as an electromagnetic wave, has a phase for the oscillation, natural light is incoherent and is a mixture of the two senses of rotation: clockwise and counterclockwise. The electron, instead, has a helicity depending on its charge if the magnetic field is static. So one sense of polarization will be resonant at the frequency of the orbit, the other will pass freely. In the orthogonal direction, one should see two lines separated by a quantity that depends only on the dynamical magnetic moment of the charge, and one that is the initial un-displaced line. Along the field direction, the two senses of polarization produce opposite senses in the electron which is resonant at the relevant frequency. Thus the separation of the components varies linearly with the magnetic field strength, as Zeeman found, and has a pattern that depends only on the orbital properties of the electron and its charge to mass ratio, e/m.

The second experiment used a remarkably simple vacuum apparatus called the Crookes tube, an evacuated long glass cylindrical container with en electrode at one end and a phosphorescent screen at the other end that acted as a second electrode but also registered the impact of the current with a visible spot. J. J. Thomson employed this device to measure the properties of the particle responsible for electric current. His tube contained a pair of metallic plates to produce a local electric field and was placed in a magnetic field whose force could be adjusted. Because the Lorentz force depends on the charge, from the electric and magnetic field, and the mass, the charge to mass ratio could be measured when the electric field was changed and the magnetic field adjusted to cancel the deviation. The value he found was almost 2,000 times larger than for the weight of the hydrogen atom (which, although neutral, had a known mass so its charge to mass ratio was inferred). This was the electron, the first elementary particle to be identified. To perform this experiment, notice that Thomson needed the force law for a current so a microscopic particle was discovered using a classical force law. It's ironic that Thomson received an honor for his work on the vortex atom and then obliterated its basis with the discovery of the electron about a decade later. This is still a daily experience for anyone who uses a standard television.[1] The deflection of the electrons is the same but the pattern is more complicated. A color television screen, rather than being a continuous film, is discretize into dots of various colors and three beams are steered together with modulated intensity of the current depending on the input signal. The same technique was also the origin of the modern oscilloscope.

The third experiment used charged oil droplets, which are good conductors and retain the charge, suspend them in an electric field against gravity under high vacuum. From the measurement of the size of the drops using a magnifier and seeing at what field they remained in equilibrium, the quantity of liberated charge could be determined. This ingenious experiment was performed by Robert Millikan (1868–1953), who also showed the discreteness of the charge in terms of a fundamental value. While the Thomson experiment yielded the charge to mass ratio for the charges in the current, the Millikan experiment directly measured the unit of charge. The measurements of the charge of the electron, and its charge to mass ratio, were based on mechanics and magneto- and electrostatics. They showed that the charge is equal to that of the ions, that a unit of charge corresponds to the atomic number already found by the chemists, but that the mass of the particle is extremely small, of order 0.1% that of the ions. This cannot account for the bulk of the mass of the atom. The discovery of charge as a particle property, along with electrodynamics, completed the classical program of applying force to field.

THE STATISTICAL MECHANICAL ROOTS OF THE QUANTUM THEORY

Maxwell and Gibbs, in their development of statistical mechanics, had successfully described how to treat the interactions between particles in a gas and, with that, to derive the gas laws. But they had been able to almost completely ignore their internal structure. In short, they had not addressed the structure of the atom. Boltzmann, who had been one of the principal contributors to the continuum theory, also provided the first model for how the internal energy is distributed. He assumed that any atom is a bound system consisting of discrete energy levels, states, whose occupation was statistically determined only by the state's energy relative to the mean thermal energy of the system and on the *statistical weight* of the level—the number of ways the individual states can be populated.[2] Spectral lines furnish the key since these are characteristic of the substance and at discrete frequencies. How the states are structured, and why they don't form a continuum, were separate questions. Transitions between states were not explicitly dealt with nor was there any explanation for how the radiation and matter couple. However, that wasn't necessary. Boltzmann was treating systems in thermal equilibrium. Since the rates for all individual processes collectively balance, it suffices to know the population distribution they generate, thus avoiding any "how" questions. An important feature of this approach, you'll notice, is how it avoids any forces. The states are determined somehow; this was proposed before the discovery of the electron so Boltzmann didn't even detail what he meant by the energy levels themselves. That would come later. It was enough that the energies are discrete. His distribution was the probability of a number of atoms in a gas being in some state relative to the lowest energy level, the ground state, and because it was an equilibrium calculation didn't treat the individual atoms, just an ensemble.

There was another connection between light and microphysics. Boltzmann had found the relation between the energy density of equilibrium radiation within a closed, hot cavity and its pressure. The interaction with the walls was a separate matter. In equilibrium, all radiation incident on the wall is either absorbed or reflected. For a blackbody radiation, the limiting case, the energy is absorbed equally at all frequencies and is in energy balance between the radiation absorbed and emitted independent of the frequency, there is a spectral energy distribution in equilibrium whose total flux depends only on temperature. Kichhoff and Bunsen had found, earlier in the nineteenth century, that the ratio of the emissivity to the absorption is a universal function only of temperature but the precise form of the spectrum was unknown. Its integrated intensity had been measured in 1879 by Jozef Stefan (1835–1893) and Boltzmann's derivation of the radiation pressure law also predicted the measured T^4 temperature dependence. There seemed to be a connection between the atomic states and this radiative process since the cavity is at a fixed temperature and the radiation emitted by the walls is due to the de-excitation of the constituent atoms. As for Boltzmann's atom, the distribution is formally universal, independent of the properties of the matter. This was the approach adopted by Max Planck (1858–1947) who, in 1900, announced the law for the energy distribution of the radiation. He was thinking about the radiation itself, combining modes in a cavity and assuming the individual oscillators would not interfere. The electromagnetic field was modeled as discrete packets of light, the *quanta*, whose energy was assumed to be an integer multiple of a fundamental quantity, the frequency times a new physical constant, h that has the dimensions of action (that is, energy times time or momentum times displacement). These oscillators only radiate by transitions between the stationary states, the energies are not the same as those of the emitted photons and only the interaction produces a change. The law agreed perfectly with experimental measurements of the spectrum of blackbody radiation but it left the nature of these quanta of action unexplained. There was another connection between light and the atom. This was the photoelectric effect, the emission of charges when a metal cathode is exposed to blue or ultraviolet light, that was discovered by Philipp Lenard (1862–1947) in 1902. There were two parts to the effect. He used a high-intensity carbon arc to illuminate a metal plate in an evacuated tube, similar to that used by Thomson, but with an anode whose potential could be changed to prevent the impact of low-energy electrons. He found that the intensity of the current was directly proportional to the intensity of the light but not the energy of the electrons, which depended instead on the wavelength of the light. Millikan extended Lenard's results and found that the constant of proportionality had the same value as Planck's constant of radiation. The photoelectric law, that the energy of the ejected electron is linearly proportional to the frequency of the light, was an unexplained but fundamental result.

This is where Einstein started in 1905. He was already thoroughly immersed in thinking about how to apply random processes to microscopic phenomena from his work on Brownian motion. Now he addressed the problem of how fluctuations would interact with a bound system. His approach was to abstract atomic structure as far as possible, reducing the system to a two level harmonic oscillator. Following Planck, he assumed that emission and absorption take place only in discrete jumps, which were subsequently called "photons." I'll now use that term to signify Planck's quanta of action. These jumps occur with some probability distribution. He distinguished two types of transitions, those that were "stimulated" by the radiation in the cavity, the rate of which are proportional to the intensity, and those that occur "spontaneously." While the stimulated processes can either add or subtract energy from the system by, respectively, absorption or emission, the spontaneous transitions can only produce emission. This was also consistent with his understanding of entropy in thermal baths. To see how this affects the radiation, we suppose an atom, with only two levels with energies E_1 and $E_2 > E_1$, is inside a box within which there is a collection of particles with which it collides. There is also radiation in the box, it's hot, and the radiation is both absorbed and reemitted by the atom. Imagine that we have not one but a slew of atoms, and the relative level populations are given by the Boltzmann distribution because the particle collision time is sufficiently short relative to any radiation absorption or emission process. We now compare the rates for only the radiative processes. That which produces absorption of a number of incident photons, with arrival rate I at the frequency corresponding to the separation of the levels, can be absorbed depending on the relative population of the lower level and the cross section for the interaction. Einstein wrote this as (*rate of absorption*) $= n_1 B(1 \rightarrow 2) I$. Here n_1 is the number of atoms in the first state relative to the total number in the box, the "population," and I is the incident or emitted intensity. He assumed the two electromagnetic distributions would be identical in equilibrium. The inverse process, when a photon produces a downward transition, is called the *stimulated rate*, (*rate of stimulated emission*) $= n_2 B(2 \rightarrow 1)I$ since it is assumed the same "type" of photon can produce a de-excitation of the upper level whose population is n_2. The B coefficients are the probabilities of the process (actually, they're the rates but the assumption was the processes are random in time).

If this were the only set of possible processes, the balance between them would depend only on the relative transition probabilities and not the intensity of the radiation. But when the upper state is excited, it can spontaneously decay, independent of the energy density of the radiation, with a rate $n_2 A(2 \rightarrow 1)$, A being the probability of a spontaneous emission in an analogous manner to B. This is a statistical process, one produced by the fluctuations in the level populations and the chance that one of these will be large enough to cause the system to relax. It is inherently unpredictable. Therefore, there is a source for new photons, one that is independent of the radiation and only comes from the populations themselves. Einstein's next step was to connect the radiation with the mechanical processes in the gas, to say the level populations are produced only because of the collisions, the rate of which depends only on the temperature, as Boltzmann had found. The ratio of the level populations then depends only on their energy difference. The radiation density and spectral distribution in the box will be precisely what are required to maintain the equilibrium between the radiation and the oscillators in the box. This was Planck's model and Einstein showed that if the transition probabilities are constant he recovered the known result for blackbody radiation. To do this, however, Einstein made a revolutionary assumption, that the energy of the radiation of frequency ν is $\Delta E = h\nu$ and corresponds to the *energy difference* of the two levels. This was the step that Planck hadn't taken. It explained the photoelectric effect and it is the "origin" of the photon. It also followed that the radiation will be independent of the level distribution of the atomic system and, because only *ratios* of the probabilities and ratios of the level populations are required there is a direct relation between the probability of absorption $B(1 \rightarrow 2)$ and emission $B(2 \rightarrow 1)$, and that of spontaneous emission $A(2 \rightarrow 1)$ that holds for every pair of levels in the atom. The almost astonishing simplicity of this argument may hide the essential conundrum. This is a relation between statistical quantities, not deterministic ones. The absorption and emission occur only with relative probabilities, not deterministic rates. The Planck distribution is then, as was the Maxwell–Boltzmann distribution, the relative chance of observing a photon *at a given energy*, not the certainty of seeing it. A purely deterministic mechanical process might produce the radiation incident on the cavity but the radiation inside is completely random.

Atomic Structure and Quantum Theory

Radioactive emission provided a new tool to probe atomic structure. This consisted of using α particles, the least penetrating of the three types emitted by radioactive substances. These carry positive charge and have the mass of a helium atom. They provided Ernest Rutherford (1871–1937) the tool for probing the structure of atoms in solids, where their fixed arrangement simplified the dynamics. The scattering experiments he performed in 1911 established that the mass of the nucleus is centrally very concentrated. The experiment was beautifully simple. A beam of α particles was aimed at a thin metal foil; Rutherford chose gold because it has a large atomic weight, hence a large positively charged constituent, and could be easily milled into very thin layers. The particles scattered in an unexpected way. Instead of a broad range of scattering angles, that was expected expect from a uniform distribution of charge, most simply passed directly through the foil with minimal deviation and only a very few backscattered. Rutherford showed that this could be explained using a classical two body calculation in which like charges interact electrostatically, one in a stationary pointlike mass and the other, lighter charge moving in hyperbolic paths. The motion was inconsistent with a uniformly mixed, neutral blob. Instead, the scattering experiments required the bound atomic electrons to be far from the nucleus. Their location would be explained if they execute bound orbits around the central charge.

Figure 9.1: Niels Bohr. Image copyright History of Science Collections, University of Oklahoma Libraries.

This solar system in miniature was the starting point for Niels Bohr's (1885–1962) idea of quantization. Bohr was a research associate with Rutherford at Manchester around the time the new atomic model was being formulated. In essence, it is a trick, a way of accounting for the stability of the orbits while at the same time permitting emission when necessary. The problem came from classical electromagnetic electron theory. A moving charge radiates and, since it's losing energy, its orbit decays continually. This would lead to an almost instantaneous collapse of matter which, as your daily existence demonstrates, doesn't happen. Instead, Bohr assumed that the orbital angular momentum is assumed to take only specific, discrete values. A proportionality constant with the right dimensions was already known for this, h, Planck's constant. He also assumed the photon picture for transitions between these levels, that they occur according to Einstein's statistical formulation of the photon, and that the change in the energy produced the emission of a pulse of light, a photon, whose frequency is given by Einstein's photoelectric relation $\Delta E = h\nu$. The force maintaining the orbits was classical, electrostatic attraction of the nucleus, and the electrons were treated as pointlike orbiting low-mass particles. The change from the classical dynamics was the introduction of a quantization rule for the levels. Bohr then assumed that the discreteness of the orbital angular momentum corresponded to a discrete radial distribution in which the distance from the nucleus is an integer multiple, n of a fundamental length, a_0, now called the "Bohr radius" (about 0.5 Å or 5×10^{-5} microns) and obtained the known hydrogen line series discovered by Balmer and Lyman and predicted the proportionality constant for the frequencies that had been determined by Rydberg in terms of the charge and mass of the electron, e and m_e, and h with no free parameters. His next success was the prediction of the spectrum of ionized helium for which the reduced mass of the orbit shifts the center of mass because of the difference in atomic weight between hydrogen and the one electron helium ion. The series, discovered in the 1890s by Edward Pickering in the optical spectra of very hot stars and at the time unobtainable in the laboratory was a tremendous success of the model. Laboratory line spectra thus became the tool to delineate the permitted orbits within this system and atomic structure could now be studied by proxy.

The fine structure of orbits was explained by a separate but related mechanism. As it did for the Newtonian problem of gravitation, the Kepler problem appeared in the atom but now in a different guise. Recall that the rest mass is not the important thing for a moving mass, its inertia depends on its motion. Thus, for an atomic orbit, the mass of the electron depends on the local motion. If the angular momentum can be distributed discretely for any energy in units of $\hbar = h/2\pi$, the eccentricity of an orbit is determined by the ratio of the binding energy to the angular kinetic energy through **L**, the total angular momentum. If we think of the orbits as ellipses, in a central potential they close. But if there are other charges, or if the mass of the electron isn't constant, the orbit will precess. If this new motion is also quantized, the amplitude in energy—the splitting of the lines—will be smaller than that produced by the angular momentum but it will be a necessary effect of the dynamics. This explanation, due to Arnold Sommerfeld (1868–1951), not only appeared successful but furthered the program of modifying the classical notion of force to the atom.

The problem of how these orbits arrange themselves stably was a different story. Bohr postulated that they group in shells with a maximum permissible number. This explained the origin of the statistical weights for the levels that Boltzmann had introduced as the irreducible number of possible combinations for particles in each atomic state, independent of temperature. But there was no explanation for how this arrangement was achieved. The clue came from helium, the simplest two electron system. Wolfgang Pauli (1900–1958) introduced the "exclusion principle" (1925) to explain the absence of some transitions predicted purely on the basis of the distribution of the atomic levels, by stating that no two electrons can occupy precisely the same state. This required introducing another, unexplained, quantum number to those required for the radial and angular motions. In an attempt to explain both the Zeeman effect and the Pauli principle, in the same year Samuel Goudschmit and George Uhlenbeck proposed that spin would account for a contribution to the magnetic reaction of the atom by endowing the electron with its own intrinsic magnetic moment. The effect of an imposed magnetic field therefore aligns these spins and splits the atomic levels to produce the Zeeman effect for the subsequent transitions. The difficulty was immediately pointed out by Lorentz, that since the electron is a pointlike mass it could not have an axial rotation. Nor was it clear why the number of components should be *half-integer* multiples of the Bohr magneton, the elementary unit of magnetization. But despite these problems, Goudschmidt and Uhlenbeck's paper was sent to *Naturwissenshaft* by their professor, Paul Ehrenfest, and subsequently an English version was published in *Nature* accompanied by a recommendation from Bohr.[3] Thus was the spirit of the times. Even crazy proposals had a chance of being correct. The explanation for the strange factor of 1/2 came surprisingly quickly. The dynamical problem of an orbiting, spinning mass was solved by L. H. Thomas (1926) in another application of relativistic mechanics to atomic structure, the motion of a spinning particle orbiting in an electrostatic field. He found a multiplicative factor of 1/2 that appeared to resolve the last quantum mystery. The solution later became the basis of the fully coherent theory of synchrotron radiation, the emission of light due to free motion of electrons in a magnetic field by Julian Schwinger but, as we will see, with the demise of the "mini-solar system" model the explanation isn't applicable to atomic structure.

Bohr was also able to treat molecular motion within the quantum theory by assuming the quantization rule for each available motion. Since molecules are linked atoms, and therefore neither spherical nor pointlike, the nuclei were treated as masses glued together by the orbiting electrons. This is actually a very strange recipe, it combined multi-center dynamics (which we saw from the three body problem leads to incommensurable orbits and ultimately to chaos) and a number of paradoxes of motion, but it preserves the force concept. Atoms in molecules behave as if tied together by springs, hence the interaction potential is imagined as a source for a spring constant from which the oscillation frequencies can be computed. For rotation, the simplest procedure was followed, to determine the moment of inertia for any geometry of the component atoms and then to quantize the rotational motions in the same manner as the orbits. The regularity of the individual transitions was accommodated by writing $E_J = \frac{1}{2I}J(J+1)\hbar$ where I is the moment of inertia and J is the rotational angular momentum, and to then use the Einstein–Bohr relation for the difference between energies of the rotator $E_{J'} - E_J = \frac{\hbar}{2I}[J'(J'+1) - J(J+1)]$ to find the frequency. The differences between the transitions consequently form a ladder. The same idea allowed the vibrational levels to become $E_{v'} - E_v = \hbar\omega_0(v' - v)$ where ω_0 is the resonance frequency and v and v' are the vibrational quantum numbers. Thus, by 1925, quantum theory had been successful was essentially a classical analogy with a number of apparently *ad hoc* rules for assigning states to the orbiting particles. It left unanswered, however, more basic questions of stability and dynamics.

QUANTIZING DYNAMICS: A NEW MECHANICS

In the quantum theoretical atom, neither the transition processes nor the relative strengths of the emission lines were explained. Bohr was treating only structure, how the electrons arrange themselves in the binding electrostatic field of the nucleus. As such, making the assumption that the states remain stationary and that each periodic (bound) motion is quantized, the model provides an explanation for the discrete levels and their relative statistical weights (their *degeneracies*) while still avoiding the details of the transitions. In effect, the picture is again an equilibrium, of sorts: if a rule is imposed that forces the states to be stationary, that the orbiting electron cannot radiate, then by imposing a discretization in the distances and angular momenta you obtain the spectral sequences (in the one electron case) and a method for inverting the observed line distributions to obtain the quantum numbers for the derived states. But *why* the atomic states should be quantized was not addressed, or indeed addressable, within the theory. That required something more, a new mechanics. From the beginning, the important difference between the classical and quantum pictures is that h, the Planck constant, has the dimensions of action and angular momentum. Although both are known from macroscopic dynamics, they are continuous properties of a system and neither has a characteristic value.

If on the scale of everyday life masses and charges follow the classical laws of motion, or their relativistic extensions, there must be a way to connect this with the atomic realm. How to make the transition was outlined by Bohr in his third paper on atomic structure in 1913. In what he called the "Correspondence Principle," he asserted that the way from classical mechanics to a microscopic system was to quantize the action in units of the Planck constant, h, and to make all derivatives discrete by taking the change in the energy between states. It was, in effect, going backwards from a smooth to discontinuous rate of change of any measurable quantity, contrary to the way that the calculus had originally been developed! For Bohr, this was plausible because the existence of spectral lines and energy levels demanded that a dynamical quantity couldn't ever change by an infinitesimal amount, these had to be discrete jumps. This required a complete change in the Newtonian idea of a force. Accelerations became impulsive changes in the momentum per unit mass, similar to collisions. Time derivatives were replaced by multiples of the frequency and, you'll recall, the frequency is a change in the energy when going between quantum states. In Hamilton's formalism of mechanics the frequency is the change in the Hamiltonian, or total energy, with respect to the action. But now, since the action is quantized, as is the energy, the photoelectric relation of frequency and change in the energy was recovered.

By the beginning of the 1920s, quantum theory was a mass of tools, guesses, and analogies in search of a foundation. Guided by Bohr's axioms, the continuous dynamical equations could be quantized by taking any changes with respect to the action to be discrete. But the difficulty was that although the orbital picture explains spectra, it fails to "save the phenomena" for matter. Building atoms is simple, a problem in discretizing the manifold of possible orbits among the interacting electrons in the electrostatic potential of the nucleus. But molecules, and more complex aggregates, were much more difficult to imagine. The orbits of the individual atoms must somehow overlap and link the constituents to form stable structures, an impossible task in so simple a system as the hydrogen molecule (although not for the ionized molecule which is a three body problem). The demise of this micro-celestial mechanics all but severed the last direct connection with the classical mechanical roots of the microscopic picture. The one model that remained was the harmonic oscillator and the principal tool was the Hamiltonian. By analogy with the Hamiltonian construction of mechanics, the change of the energy with respect to the action is the frequency. Then it follows, by construction of the quantized system, that the difference in the energy between two levels is the energy of the emitted quantum so any changes in the system the time also something occurs by jumps.

Wave Mechanics

Wave mechanics, the first step in the transformation of the force concept in the microphysics, was initiated by Louis de Broglie (1875–1960) when he showed how a particle can be *treated* like a wave, an idea suggested by the comparison of the classical and quantized picture for electromagnetic emission, in which a mass m having momentum p appears to behave like a wave with a wavelength $\lambda = h/p$. At about the same time, Arthur Compton (1892–1962) had sought a description of photon scattering by free photons assuming a nonrelativistic particle picture. Let's digress for a moment to see how that works.

A parcel of light carries a momentum $p = E/c = \hbar\omega/c$. If an electron at rest is hit by a photon, the light scatters. Since the momentum of the photon is along its propagation direction, assuming the scattering takes place at an arbitrary angle, the electron gains energy at the photon's expense. Energy and momentum are conserved in this collision. But, and this is the most important point, a change in the photon's momentum (and energy) means its *frequency* also changes: increasing the momentum of the electron produces a downshift in the photon's frequency because the parcel of light is massless and, as observed by Compton in 1922, X-ray photons lose energy while the electrons gain momentum. A new feature is that the collision has a threshold at characteristic wavelength that depends only on the particle's rest mass, $\lambda_C = h/(m_0c)$, that is now called the Compton wavelength. Alternatively, we can say the process is nearly resonant when the photon energy is equal to the particle's rest energy, so the governing parameter is $h\omega/(m_0c^2)$ which, when equal to unity, gives the Compton wavelength. The advantage of this approach is how it avoids any questions about *how* the interaction takes place. By using a simple ballistic collision, the course of the particle and photon can be traced to and from the collision without introducing any model for the interaction. By this straightforward calculation, made revolutionary by its explicitly including photons as "particles of light," a striking result had emerged: the energy of a photon changes after it hits a stationary electron but the change depends on the angle between the incident and final directions of the scattered photon. The electron is accelerated and the energy and momentum behave for this system precisely as if the photon is a relativistic particle with momentum but without a rest mass.

A similar picture was exploited by de Broglie in his doctoral thesis (1923–1924) at Paris under the direction of Langevin, although he knew neither Compton's result nor its explanation. De Broglie knew that light is classically well described as a wave phenomenon. The controversy of the first part of the nineteenth century about whether it is a particle or wave had long before been, apparently resolved experimentally and theoretically in favor of the wave picture. The success of the photon changed this consensus. Instead of forcing a choice, de Broglie accepted this duality and asked what would be the consequence of extending this to massive particles, e.g., the electron. Within the Bohr atom, although exploiting the quantum numerology to derive orbits, the origin of the stationarity of the states was unexplained; it was imposed by construction. If instead the electron could be wavelike, this discreteness could be naturally explained as a standing wave. At certain, fixed distances from the nucleus a bound electron could, if it is a wave, interfere unless the state is a harmonic. Thus it would be a natural explanation for the structure of matter. The characteristic wavelength he computed depends only on the particle's momentum, p, $\lambda_B = h/p$. Since the momentum of a bound orbiting electron is related to its energy by the virial theorem, this yielded the corresponding states. The success of this picture, along with a relativistic justification contained in the thesis and the first of a series of papers resulting from it, made this a serious contending model for matter, albeit a very strange one. It had, however, some particularly attractive features. The model naturally exploited the phase space of a particle. Recall that h has the dimension of action and that h^3 is the same as an elementary volume of phase space. Thus, you could say, if there is only one particles per unit in phase space, for some reason, then you would expect each to have an equivalent "width" of the de Broglie wavelength for a momentum.

Langevin sent a copy of de Broglie's thesis to Peter Debye (1884–1966) at the ETH in Zürich and to Einstein at Berlin. Both read it with interest. Einstein's reaction was supportive and spurred further interest in the idea both at Paris and elsewhere, such was the importance of his imprimatur. In Zürich Debye asked his colleague, Erwin Schrödinger (1887–1961), to take a look at the work and report on it at the group seminar a few weeks later. This turned out to be an auspicious request. Although at first reluctant to consider it, Schrödinger accepted the task and in 1926 published his solution to the hydrogen atom and its generalization to atomic mechanics. He examined how this wave picture is indeed applicable to particle motion in a bound system by using the Hamiltonian function. He used similar reasoning to the approach Hamilton had introduced with his *eikonal* function.[4] Schrödinger realized that he could write a *wave function* whose exponent is the eikonal, or action, and that this would furnish the equation for a propagating wave. At first he tried a relativistic formalism and derived a second order differential equation of a wave type with a potential, the precise form having already been examined by Helmholtz in the context of acoustics. But for the nonrelativistic case the situation was different. The resulting equation is a *diffusion* type, first order in time and second order in space with the energy as a proper value, or *eigenvalue*, of the stationary states. For a freely moving particle, this yields a gaussian function. For a stationary particle in an electrostatic field, produced for instance by the nucleus, the eigenvalue spectrum is the same as the distribution of states in the hydrogen atom: the de Broglie analogy yielded the Bohr states.

The technical mathematical problem of how to solve the type of differential equation of which Schrödinger's is an example had been well studied for nearly a century, starting with Liouville in the middle of the nineteenth century. General solutions were known for diffusion and wave equations with both initial conditions and fixed boundaries and Schrödinger was thoroughly familiar with that work. The equation he derived had, however, a special feature. The energy of the state is the solution and, depending on the coordinate system, takes specific values that depend on integers. In other words, the solutions are *quantized* and this was the most remarkable feature of the equation. The solution for the hydrogen atom, a centrally symmetric electrostatic potential with a single electron, separates into three components, one radial and two angular. The solutions to the angular equations are a classes of function found by Legendre, while those in radius are functions that had been found by Hermite, both long before Schrödinger. To obtain the angle-dependent solutions he identified the integer with the angular momentum, for the radial it was with the principle quantum number. In so doing, he recovered the dependence of the energy on the two quantum numbers that had been discovered from the Bohr atom and spectroscopic analysis. It was now possible to compute atomic structure *without orbits*. But at the same time it was far from clear what these wave functions meant because they were not the thing measured, being instead the means for computing the energies by taking averages over the volume. By the correspondence principle, any average is the same as the classically measurable quantity (for instance, just as the mean of the Hamiltonian, $\langle H \rangle$, is the measurable energy, the mean momentum, $\langle p \rangle$, is the measured value for the distribution and that corresponds to the classical momentum). The dynamical equations of classical mechanics could be recovered, although it required a change in the normalization to remove Planck's constant (which is the single critical departure from the classical result). Perturbations could also be handled, and the change in the energies of atomic levels when an external electric (Stark effect) or magnetic (Zeeman effect) is present could be computed using the same formalism. Also coincidences between levels, the *degeneracy* or statistical weight, followed naturally from the wave function representation. The machinery developed quickly in Schrödinger's hands, the entire foundation was laid within one year.

From Matrix Mechanics to Quantum Mechanics

At the same time, an alternative view of atomic dynamics was being developed in the Göttingen circle centered around the physicist Max Born (1882–1970) and the mathematician David Hilbert (1862–1943). Their approach was more closely linked to the Bohr program of quantizing dynamics of particles and orbits by using operator methods. This was a route taken by Werner Heisenberg (1901–1976). Although his thesis work with Sommerfeld was on fluid turbulence, he was drawn to microphysical problems and on Sommerfeld's recommendation joined the Göttingen group. His starting point, unlike de Broglie, was to make the Correspondence Principle the central axiom of an effort to discretize mechanics. In his paper of 1925, "Quantum-mechanical re-interpretation of kinematic and mechanical relations," Heisenberg replaced the orbits with a harmonic oscillator and wrote the change in the momentum according to Bohr's principle, replacing the position as a function of time with a set of discrete, quantized, amplitudes of oscillations whose frequencies depend on the change in the quantum number. By this substitution, he was able to solve for the energy of the oscillator in terms of the quantum state which he found to be an integer multiple of h times the fundamental frequency. But he was faced with another very strange result for the position and momentum. For a classical phase space it doesn't matter if we take the produce xp or px but Heisenberg found that $xp - px = i\hbar$, with $i = \sqrt{} = 1$. The two variables didn't commute. Although such algebras had been known since Grassmann and Hamilton—indeed, they had been the point of departure for vectors and quaternions—nothing like this was expected from a mechanical system of operators. Heisenberg went on to show that the energy has a nonvanishing offset; even in its ground state such a simple system would never have zero energy. In a last step, he showed that this could be extended to a nonharmonic oscillator, one with a small cubic term that mimicked a perturbation. He immediately presented the result to Born, who recognized it as the algebra of noncommuting matrices. Together with Pascual Jordan (1902–1980), Born derived the equations of motion and in less than three months after the submission of Heisenberg's paper they published "On Quantum Mechanics." The third paper in this series, connecting with atomic structure, was published in 1926 along with Heisenberg (often called the "Dreimännerarbeit" or "Three Men's Work") in which they derived the equations of motion for a generalized potential law and quantum mechanics was born.

Why a harmonic oscillator furnishes the basic model can be understood by thinking again about orbits. An circular orbit has a minimum angular momentum for a given energy so an increase in the angular momentum produces an elliptical orbit. The particle oscillates radially in the potential field around the mean distance given by its binding energy. Thus, this looks like a radial periodic oscillation excited with respect to a reference circular orbit that is the minimum state. The oscillation has a characteristic frequency that depends on the potential, hence the particle's kinetic energy, because by the virial theorem the kinetic energy is strictly tied to the potential energy for a bound system. A departure from a simple $1/r$ potential leads to anharmonic behavior which explains why Heisenberg treated the cubic perturbation. The classical result of Jacobi and Clausius transforms into this new framework directly by construction. If we now "quantize the motions"—that is, we assume both the commutation laws and assume all angular momenta are integer multiples of $\hbar$—then the oscillators are stable (they don't radiate, again by construction) and their energies are fixed only by the combination of their quantum numbers. Thus we arrive at a picture of oscillators similar to that use by Planck in his derivation of the radiation law but with very different consequences. Heisenberg used this to show that the spectrum is a linear relation between the energy and the principal quantum number. Imagine a mass painted with a phosphorescent coating attached to a spring. Let the mass oscillate and, in a dark room, take a time exposure of the motion. The image will show the ends as more intense since the motion is slower at the turning points so the times are unequal. If pulled with a greater force the maxima are displaced depending on the energy (the extension) of the spring). Now for a sufficient extension the oscillation becomes anharmonic. This is the same as a probabilistic model: if we ask where there is a chance, on casual observation, of finding the mass it will generally be closest to the end. This looks just like an orbit where if we take a photo we will see the same projected image.

The resolution of the mysterious commutation relation at the base of the new mechanics came again from Heisenberg in 1926. He showed that this noncommutativity comes from how we know the phase space coordinates of a particle. He showed that there is an intrinsic, irreducible fuzziness to any particle. In what he called the "uncertainty principle," the more accurately you know one of the phase space coordinates the less accurately you know the other, showing the

$\Delta x \Delta p = \hbar$. These are the conjugate variables from dynamics and by changing energy and time he found that the result could also be extended to $\Delta E \Delta t = \hbar$. This meant that if you have a state with a finite lifetime, for any reason, the shorter the lifetime the broader the energy level. It also applied to particle motion. It also accounted for particle decay since the rest energy is then inversely related to the decay time. This meant that a fluctuation in any of the conjugate variables determines that in the other. For the momentum, this naturally explained why a particle could have a jitter. But this also provided at least a clue to solving the first conceptual problem, that of the jumps. Because of fluctuations in its energy, a stationary state also has a finite lifetime. The increased widths in the lines arising from successively higher transitions in hydrogen that gave rise to the original orbital designations could now be explained.[5] The epistemological price seemed to be high, it isn't possible to state precisely, with perfect accuracy, where a particle is if you have a measurement of its momentum. Heisenberg outlined a thought experiment in a series of lectures in 1926 at Chicago that provided a way to look at this. He imagined a microscope (with γ-rays as the illuminating photons to insure the Compton effect) such that any measurement made on the position of the particle inevitably and unpredictably disturbs its state of motion. Further elaborated by Bohr, this became known as the *Copenhagen interpretation*.

For the concept of forces, this removed the last support for the notion of absolute space and time. For macroscopic bodies this uncertainty doesn't matter, that's why Newtonian and relativistic mechanics work for things on a large scale. But when it comes to how matter behaves at the level that requires quantum mechanics, the only statements we can make are about probabilities of things happening: only the distribution of values can be predicted for measurements, not specific values. Bohr added an interpretative framework to help resolve this confusion. In his Como lecture of 1927, he proposed a sort of philosophical compromise, the principle of "complementarity." He proposed that since the two treatments of quantum mechanics were compatible and were formally equivalent, as Schrödinger had shown, the contrasting picture of particles and waves should be accepted as working hypotheses. Much as Newton had proposed for gravitation, without knowing the "why" the mechanical predictions were sufficient to proceed along parallel lines. Internal atomic dynamics was replaced with jumps and stationary states with transitions being time dependent interactions through fields. Dynamics would be represented by wave packets with the mass and charge of a particle a measurable quantity but accepting the intrinsic uncertainty connected with the disturbance produced in its phase space position because of uncontrollable interactions between the measuring device and the thing being measured.

An even closer connection with statistical representations came from a different direction. Shortly after the publication of Schrödinger's first papers, including the time dependence, Born realized the link between the wave function and probability theory and through this, introduced a fundamentally new interpretation of the quantum theoretical picture. From the time dependent Schrödinger equation, he found that the *square* of the wave function had the same properties as a *probability density* in space, the chance to finding a particle at some point. This moves with a center of mass momentum $\langle p \rangle$ and obeys the continuity equation. He therefore attributed this to the intrinsic "fuzziness" of the particle. This distribution follows immediately from the wave function but also carries an important feature. The amplitudes of two waves, in classical physics, interfere when superimposed. In Maxwell's ether, this finds a natural analog in the superposition of two motions for the elastic medium. In the quantum mechanical picture, this requires a change in the description. Instead we talk about correlations. But more to the point, two particles can now appear to interfere with each other! Each amplitude has an associated phase and, when we co-add particles, we are adding them coherently in phase. The motion depends on the potential, the particle itself only statistically responds to a "force." This was highlighted by a thought experiment by Aharanov and Bohm (1959) that has since been verified experimentally. A beam of electrons is divided equally into two parallel paths that pass a current carrying solenoid. The magnetic field is confined within the coil, there is no gradient in the external space. But the two beams pass on opposite sides of the coil and, therefore, experiences phase shifts of equal magnitude and opposite sign. When they recombine they interfere and produce fringes where the electrons are either correlated or anticorrelated despite the absence of a force. The potential alone is sufficient to account for quantum mechanical measurements.

At around the same time, at Cambridge, Paul A. M. Dirac (1902–1984) had been told about Heisenberg's work by R. H. Fowler, one of the few English physicists who had both followed and understood the continental developments. Thinking about the commutation relations, Dirac made an important connection, that they were analogous to a mathematical device known from Hamiltonian mechanics. When transforming between coordinate representations, called a *contact transformation*, the change is expressed using the Poisson bracket, which had also appeared in Lie's transformation theory (see the appendix for a brief discussion). It's a striking example of the generality of the canonical coordinate representation that the same basic equations appear in such diverse cases. They were most familiar from celestial mechanics where they had extensively used to change between spatial coordinates to the more convenient, and more easily measured, orbital parameters. For Dirac, the commutation rules became the fundamental axioms of the new mechanics, much as vectors and quaternions had almost a century earlier. He distinguished two types of numbers, so-called *c-numbers*, quantities that commute (classical), and *q-numbers*, those (quantum) that do not. He realized that the canonical coordinate pairs were q-numbers while the measurables, eigenvalues, must be c-numbers. He also introduced a beautifully simple, general notation for the states, the *bracket*, that at once corresponds to the statistical averaging over states and expresses the properties of the operators. A *ket*, $|n\rangle$, is the state, a *bra* is its complex conjugate, $\langle n|$, and the expectation value is then $\langle n|Q|n\rangle$ for any Hermitian operator Q.[6] A similar notation had been used for the scalar product of two vectors $\langle u, v\rangle$ where it is commutative, i.e., $\langle u, v\rangle = \langle v, u\rangle$ for real (u, v); it had also been used for Fourier transforms, with which one passes between canonically conjugate quantities, and for algebraic operations on generalized functions, The two quantities are called orthogonal when $\langle u, v\rangle = 0$ so similarly Dirac imposed the rule that two uncorrelated states have $\langle u|v\rangle = 0$ (Hilbert and other mathematicians had already used this sort of notation for taking scalar products, so its extension was natural to quantum operations).

Dirac extended the concept of "state." By analogy with photons, he took the principle of superposition as his central postulate along with the inherent indeterminacy of measurements. A state can be decomposed into a linear superposition of other states, the correlations of which are the scalar products or, as Dirac wrote:

> The intermediate character of the state formed by superposition thus expresses itself through the probability of a particular result for an observation being intermediate between the corresponding probabilities for the original states.

In his *The Principles of Quantum Mechanics* (first published in 1930 and still in print today) Dirac defined a fundamental shift in viewpoint in contrasting the algebraic and continuous approaches, writing:

> The assumption of superposition relationships between the states leads to a mathematical theory in which the equations that define are linear in the unknowns. In consequence of this, people have tried to establish analogies with systems in classical mechanics, such as vibrating strings or membranes, which are governed by linear equations and for which, therefore, a superposition principle holds. Such analogies have led to the name 'Wave Mechanics' being sometimes given to quantum mechanics. It is important to remember, however, that *the superposition that occurs in quantum mechanics is of an essentially different nature from any occurring in classical theory*, as is shown by the fact that the quantum superposition principle demands indeterminacy in the results of observations in order to be capable of a sensible physical interpretation. The analogies are likely to be misleading.

He realized that in Hamiltonian mechanics there is a special relationship between the dynamical variables, the Poisson bracket (used extensively for coordinate transformations in celestial mechanics).[7] Dirac then showed through manipulation of the brackets that for any arbitrary pair of quantities, $uv - vu =$ (constant) $\times [u, v]$ and therefore was able to arrive at Heisenberg's relations $qp - pq = i\hbar$, not in terms of derivatives but the operators themselves, and identified the constant factor as an imaginary number. This last result he called the *fundamental quantum conditions*. He then introduced a way of speaking that has since come to be dominant in the description of quantum mechanics, that when $\hbar$ becomes vanishingly small we recover classical mechanics. This is, perhaps, one of the strangest features of the theory because it is saying, instead, that when the scale of the system becomes incomparably large with respect to the elementary unit of quantum phase space (when $V/\hbar^3 \gg 1$, we return to the classical world of macroscopic bodies. An amazing feature of this algebraic approach, aside from the fact that it worked, was its generality. The Göttingen formalism required considerably more manipulation of symbolic quantities and could not immediately connect the operators with observables. Instead, by using the Poisson bracket as the central operator, Dirac could immediately establish a broad range of dynamical relations. Most important, perhaps, was his realization that because the Poisson bracket in the quantum theory has the same dimensions as angular momentum—always the same ambiguity in dimension with the action—it was possible to extend the commutation relations to something that *looked* like angular momentum as an operation *without the need to think about orbits.*

Since the Bohr approach had been successful for molecular spectra it was obligatory that the new dynamics should also work. A diatomic molecule, or any molecule consisting of as many atoms as you would like, has several modes that don't require any fundamental change in the idea of an elastic force law that meterly takes discrete values for the energy. The vibrational modes of, say, a diatomic molecule have the same frequencies as those given for any harmonic oscillator bound by any potential; under small displacement the frequency is given by the second derivative of the potential (the rate of change of the force with displacement) and is a linear multiple of the vibrational quantum number, v, $E_v = \hbar\omega_0(v + \frac{1}{2})$; the fundamental frequency is ω_0, much as would hold for a spring. The extra factor is the zero point energy, the same one found by Heisenberg. The rotating molecule is almost as simple but lacks the zero point shift; its moments of inertia, I, are identical around two axes vanishes along the internuclear axis, so if the angular momentum J is quantized by a discrete number j, such that $J^2 = j(j+1)\,\hbar$ then for each rotational state has an energy $E_J = j(j+1)\,\hbar/I$.[8]

The Least Action Principle and Sums over Paths

The last reformulation for quantum mechanics, by Richard Feynman (1918–1988), appearing in 1948 as "Space-time Approach to Nonrelativistic Quantum Mechanics," finally exploited the full power of the probability. Feynman took the wave function literally in making its phase and amplitude properties of a particle, working in Dirac's interpretation of quantum states. Using Huygens' principle from wave motion, he knew that the change in a wave on interaction is a variation in its amplitude *and* phase. The new feature is to separate the interval of free motion of a particle from the moment when it interacts with another particle (a collision) or with a field. The motion is reconceived as a *propagator*, an extension of the construction you've already seen several times in these most recent chapters: remember how parallel transport supplied the tools needed for general relativity, invariant transformation groups of linear motions led to special relativity, and the eikonal was fundamental to Hamiltonian dynamics and the Hamilton–Jacobi theory? The idea is this. Take a wave with an amplitude A and phase ϕ such that:

$$\psi = Ae^{i\phi}.$$

It doesn't matter if A is real or complex, the separation of the phase is a convenience (as Hamilton implicitly did). Then the squared modulus, $|A|^2$, is independent of ϕ. Now we propagate the *amplitudes*—not the modulus—as we would for a wave by saying that

> [The amplitude at position (x, t)] = [How you get from $(x', t') \rightarrow (x, t)$] times [The amplitude at position (x, t)] summed over all possible paths

or, in symbolic form:

$$A(x,t) = \int_{\text{(all paths)}} K(x,t\,|x',t')A(x',t')\mathcal{D}(x',t'),$$

which is now called the *path integral*. This change in notation is central to the idea, that $\mathcal{D}$ is the step along any possible path, not any one in particular. At this stage the change doesn't seem that important since you don't know *which* trajectory to follow. But, by asserting that the particle "chooses" the path that minimizes the action, Feynman had the answer. The phase *is* the action, as it is for Hamilton's optical eikonal, and since that is quantized, so is the motion. This form of the transport equation was already known from potential theory and had been used for electromagnetic wave propagation, especially during the World War II in the design of waveguides and antennas for radar. The function K is a Green function (yes, George Green again!). It can also be written in relativistic form by using the four-momentum the momentum that includes the energy; the formalism is relativistically covariant, which was the crucial step in Feynman's search for a general formulation of quantum electrodynamics. We can work in four-momentum and then transform back, if necessary, to spacetime. If the change in time of the amplitude is an infinite series because (as Zeno would have said) to go from (x',t') to (x,t) you must pass through an intermediate $(x'',t')'$, and so on. Feynman realized that the free propagator is actually the first term and that the series (which had been studied extensively in the first half of the twentieth century in connection with the theory of integral equations and functionals by Fredholm and Volterra) is the sum over all possible interactions (paths) that the particle wave function ψ can follow. This, when expressed by vertices for the interactions, introduces the potentials that replace forces. The first result was an independent derivation of the Schrödinger equation. Along with the formalism, which had also been explored by Julian Schwinger, Feynman proposed a graphical technique for representing the integrals, now known as *Feynman diagrams*. These are, essentially, cartoon vector diagrams of the various steps in the propagation. But they are also calculational tools since a set of rules drawn from the formalism tell how the interpret individual parts of the graphs. Each corresponds to a precise mathematical equivalent in the developed perturbation series, the diagrams are merely pneumonics for the analytical development. But the tool is a remarkably general and valuable one, allowing the visualization of the interactions in a way Newton could have appreciated. In this way forces become fields only and are carried impulsively at interactions by particles. For instance, a propagator is a simple line. But there can be interactions and these are drawn as vertices. The cartoons are just tools, but they enormously simplified the mass of formalism by allowing the potentials and their effects to be visualized.

In this new world of trajectories without forces we still have energy and fields, the two remaining—but altered—classically defined concepts. Instead of having point particles, we trace the paths of the centers of mass as a propagation of an amplitude. Taking seriously the duality of matter means the wave function is an amplitude of a probability density. In this case, we can include all of the mechanical effects without ever dealing with forces. The path of a particle becomes a variable quantity and, using the action and the variational principles it supports, the minimum path within a field can be determined. Then this becomes the "most probable" path for a body to follow. There are no forces because we now need only the potential, a continuous field through which a body moves. For a field, the position isn't the important variable but the intensity of the field.

A particle arriving, for instance, at a slit isn't merely a point mass. The event has an associated probability of occurring. The slit, analogously to a wave source, acts as a boundary condition on the motion of the particle. Because of the uncertainty of the particles momentum and location, there is a chance that it emerges from someplace within the slit rather than the center and, as with a wave, we now have the momentum (the direction of propagation) distributed with some probability through the space after the slit. This had been experimentally verified by Davisson and Germer in 1927 as a test of de Broglie's hypothesis. To describe the subsequent motion we consider the range of all possible trajectories and compute their relative likelihood, from which we get the most direct path between two points given the constraint of the energy of the particle and the potential through which it is moved. For a classical particle, as we saw for statistical mechanics, we are dealing with a point mass and the probability is computed by asking what paths the *particle* takes. For the quantum mechanical case, we ask instead how a wave is propagated and then sum the amplitude at each point around that central path to find, through the quantum conditions, a distribution. In this way, for instance, a double slit produces a diffraction pattern for the sum of many particles that pass through it.

Feynman showed that the path integral yields the Schrödinger equation for non-relativistic motion and then extended the approach to interaction of relativistic particles and fields. For the nonrelativistic case it is the probabilistic dynamics of Brownian motion. Of all paths a particle can take between two points in a spacetime, or a phase space, one is special—the path that minimizes the action. The beginning of the path integral approach was Feynman's attempt to understand a remark by Dirac that the phase of a wave function is related to the action. In fact, this was already embedded in the mechanics of Hamilton applied to geometrical optics. Imagine a plane wave. At each point in the wave there's a phase and the amplitude propagates in space and time at constant phase (unless the wave disperses). The same can be thought of a particle in this new quantum world. In that case, the past that minimizes the change in the phase is the same one that obeys the Euler-Lagrange equations, it minimizes the action.

THE MICROSCOPIC FORCES

Quantum mechanics was able to address atomic and molecular structure and the interaction of radiation with matter using only electromagnetic potentials. But what holds the nucleus together? This was a tremendous difficulty after the discovery of the "nuclear atom." Once it was clear that the nucleus is composed of a bunch of identical positively charged massive particles, protons, their mutually repulsive potential, the Coulomb interaction, must be balanced by something. The discovery of the neutron in 1932 by James Chadwick showed the way toward a resolution of the paradox. This neutral particle has almost the same mass as the proton and nearly always the same number of particles in the nucleus.[9] The close match between the masses led Heisenberg to postulate that the two are actually states of a single particle, a nucleon, with the proton being the ground state. He and Ettore Majorana (1906–1938?) introduced a new sort of angular momentum, called *isospin*, that behaves similarly to the spin of the electron and proton and, by analogy, called one particle the "spin up" state and the other "the spin down." In the same way that scalar products of the spin are responsible for a force between electrons, they introduced an interaction that is the scalar product of the isospins. This was the first time a fundamentally new force had been proposed, the strong force, based on a phenomenological necessity.

But if these two particles are states they should be able to transform into each other like the spontaneous transitions among atomic states. This was clear from the decay of the neutron into a proton. In the process, an electron is emitted that accounts for both the mass difference and the charge balance. But when the sums were done the balance sheet was deficient, the electron—which was identified quickly with the β particle of radioactivity—didn't emerge at a single energy. Cloud chamber and ionization studies showed, instead, that it displayed a continuous rage of energies. An explanation was tentatively proposed by Pauli: the decay involves the emission of another, neutral particle of low mass and the energy is distributed among the decay products randomly and continuously. Enrico Fermi (1901–1954) jokingly dubbed this "neutrino," the little neutral thing in contrast to the neutron (in Italian, "neutrone" or big neutral particle) and in 1934 produced a theoretical explanation for the decay using creation and annihilation operators and a pointlike interaction based on an extension of quantum electrodynamics. This was the last new force, the weak interaction that holds individual particles together and dominates the electromagnetic forces on the scales even smaller than the nucleus. The particle was finally detected directly only in the 1950s but its basic properties were already scoped out from the elaborations of the Fermi theory of β-decay.

The strong force, however, did not seem to be pointlike. Instead, it required a range that was given by the size of the nucleus and the absence of its effects at any greater distance. An explanation was proposed by H. Yukawa in 1935 using the uncertainty principle. He postulated a virtual particle that is exchanged between the nucleons, independent of their electric charge. By virtual, he meant that the exchange occurs within the uncertainty time based on the mass of the hypothetical particle. This required the *mesotron* or *meson* to be intermediate mass, nearly that of the nucleons and far greater than the electron, because of which the strong force would look like a pointlike interaction with an exponential cutoff, so he could write $V(r) = [\exp(-r/L)]/r$ where L was the Compton wavelength of the meson. This is the origin of the charge independence of many nuclear properties.

The stage was set for the next unification, finally achieved in the mid-1960s by Steven Weinberg, Abdus Salam, and Sheldon Glashow. They assumed instead that the interaction is not pointlike and derived masses for the carrier significantly greater than that of the proton, about 80 GeV/c^2. The weak force is due to two particles, Z^0 and W (which come sin charged and neutral varieties), each with about the same mass. A proposal along this line had been made by Klein, on analogy with the Yukawa construction, to avoid the pointlike interaction required for the Fermi theory. But the particle mass had to be quite large, much greater than the proton's, because the range is so short and this was proposed in the extension of the idea along with the charge differences among the proposed carriers of the weak force. The photon becomes the third member of this family, now written as (γ, Z^0, $W^{\pm,0}$), a step that unifies the electromagnetic and weak forces as the *electroweak* interaction. Notice we are still using the *word* force but we are describing the interactions by dynamical potentials. The unification of the baryons and mesons was achieved by Gell Mann, Ne'eman, and Zweig, who showed that the zoo of massive particles, the *baryons*, can be reduced to states of a set of only three elementary particles, called *quarks*. To this Glashow soon added a fourth. These control the strong force through their interaction by a set of spinless particles, called *gluons*. But to go deeper into the developing picture, which is still far from complete, would take us too far afield—you will have to consult other books in this series—and here we end our journey.

NOTES

1. Flat screen displays use the polarization effect of electric fields on liquid crystals, a semi-ordered state of matter that polarize and change their optical properties depending on the local potentials.

2. You can think of a stadium in which the higher levels for the seats require more energy to climb the steps and are less populated. The lower levels are filled first and the number of seats in each row determines the populations. If there are random exchanges between the rows that depend only on a global parameter, the energy, then the rows will be peopled depending only on the mean energy of the arriving spectators.

3. The story of this paper may be an inspiration to you as a fledgling phycist. Recalling the event in 1971, Goudsmit recounted: "We had just written a short article in German and given to Ehrenfest, who wanted to send it to "Naturwissenschaften." Now it is being told that Uhlenbeck got frightened, went to Ehrenfest and said: "Don't send it off, because it probably is wrong; it is impossible, one cannot have an electron that rotates at such high speed and has the right moment." And Ehrenfest replied: "It is too late, I have sent it off already." But I do not remember the event; I never had the idea that it was wrong because I did not know enough. The one thing I remember is that Ehrenfest said to me: "Well, that is a nice idea, though it may be wrong. But you don't yet have a reputation, so you have nothing to lose." That is the only thing I remember" [from *Foundations of Modern EPR*, ed. G.R. Eaton, S.S. Eaton and K.M. Salikhov (World Scientific: Singapore, 1998)].

4. The Hamilton–Jacobi equation had been extensively discussed in the Munich circle and included as a central tool within quantum theory in Sommerfeld's book on *Atomic Structure and Spectral Lines*. Its several editions made the technique widely known. Within the equations, the individual momenta separate. They are represented by gradients of the eikonal and therefore add like phases.

5. Laboratory studies had already demonstrated that the successively lower energy series of spectral lines have increasing width. This was expressed, e.g., as s, sharp, for the first (Lyman) series, d diffuse, for the Paschen series. For the Balmer (principal or p series) and the Brackett (f, fine), the nomenclature is a bit different. These also were used to designate the orbital states for the Bohr atom.

6. A Hermitian matrix, or operator, is one for which the eigenvalues are real. These correspond, in quantum mechanics, to the ensemble averages and are the physically measurable quantities.

7. Defined by:

$$[u, v] = \sum_j \left(\frac{\partial u}{\partial q_j} \frac{\partial v}{\partial p_j} - \frac{\partial v}{\partial q_j} \frac{\partial u}{\partial p_j} \right),$$

where the sum is over the set of j coordinates appropriate to the phase space with positions q and momenta p (three dimensional, for instance). Immediately, for arbitrary (u, v) we have anticommutivity, $[u, v] = -[v, u]$, and linearity, $[u_1 + u_2, v] = [u_1, v] + [u_2, v]$.

8. The only odd feature is the discretization required for the J value but that comes from the property of the functions that satisfy the Schrödinger equation (the eigenfunctions are spherical harmonics and the associated eigenvalues are indexed by integers $j(j+1)$). This and the vibrational quantization actually had already been found from the old quantum theory and it are the only representations that survived the transition to the new mechanics. From these energies and the statistical weights of the states the populations from collisions—the Boltzmann distribution—can be computed, thus the internal state of any particles can be specified as a function of temperature.

9. An interesting point is that the nuclei of the chemical elements have the number of protons, Z—which is the same as the number of electrons and equals the atomic number—as neutrons N. An important experimental result is that there is a range allowed for the difference $N - Z$ that is neither symmetric in neutron or proton richness nor ever very large. In an analogy with thermodynamics, at some point the forces saturate and the excess particles evaporate from the binding potential.

APPENDIX: SOME MATHEMATICAL IDEAS

It is interesting thus to follow the intellectual truths of analysis in the phenomena of nature. This correspondence, of which the system of the world will offer us numerous examples, makes one of the greatest charms attached to mathematical speculations.

—P. S. Laplace, *Exposition du Système du Monde* (1799)

Many of the things you've seen in this book require mathematics you may not yet have encountered in your studies. Don't worry, you will. This appendix isn't meant to replace either formal study or serve as a textbook on the mathematical methods. Instead, much of modern mathematics has roots in questions related to physics, especially dynamics, which can often be hidden in the formalism of the abstractions. I'm hoping you'll read this appendix and use it as a sort of essay to set some of these ideas and methods in context. Although this book is about physics, the development of mathematical techniques in the seventeenth and eighteenth century is an essential part of the story since many of its sources lay in physical—often mechanical—problems. Nowhere is this more evident than the development of the calculus.

Motion is either continuous or impulsive, and we have seen the difficulties the latter presented to the medievals. They spent considerable ink on whether it was possible, or reasonable, to divide any interval into infinitely many smaller ones. For Zeno, the pre-Socratic philosopher whose paradoxes of motion were actually considerations of the infinite and infinitesimal, it was impossible to imagine the subdivided interval because an infinite number of intervals would require an infinite time to cross them all. Archimedes was able to address the problem with a first pass at the concept of a limit, and Cavalieri, Galileo, Newton, and Leibniz were finally able to exploit it to describe extension and motion.

FUNCTIONS AND SPACES

For the purpose of mechanics, we can take a trajectory as an example of a function. Speaking classically, in the language Newton, Lagrange, Laplace, and their contemporaries found familiar, t is an ordered set of coordinates in time. If a force is either smoothly varying or constant, the curve is at least continuous in its second-order changes, those that compare where the particle was at some time $t - 2\Delta t$ with now t. The velocity is also continuous to the same order because to go from $x(t - 2\Delta t)$ to $x(t)$ you need to know the velocity at the two previous times that take you from $x(t - 2\Delta t)$ to $x(t - \Delta t)$ and then to $x(t)$. Making this more precise required almost a century's investigation, culminating with Cauchy's definition of continuity. Yes, you can think of this as a relation between two variables, one independent and the other dependent, or as a track on a graph. Within mechanics the change from continuous to discontinuous motion, the importance of collisions, forced a generalization of the concept of function beyond this original form of continuity. This is connected with the way change, differentiation, is defined as we'll discuss. But if you've already seen calculus and know that the tangent to a line is the derivative of the function represented by the graph, you can see one way this can happen. In a collision, the change in speed and direction for a particle is seemingly instantaneous. It's true that for real-world cases such as automobile collisions, this isn't strictly true. But it is close enough, for a physicist, because we can make the interval during the collision as short as we'd like and examine only the state going in compared to the state going out of the site of the accident.

THE CALCULUS

Think of a small piece of a path, imagining it in one dimension, with a length L. If this is crossed in a time interval T then the speed is, of course, L/T, right? Not precisely. As you saw from the Merton rule, the speed *may* vary during T and the time interval can itself become very short. You can imagine, then, an interval Δt in which the speed is constant, but in that case the distance traversed will be $\Delta x = V\,\Delta t$ only during the interval. If we imagine V changing, then at a time $t + \Delta t$ it will be $V(t + \Delta t)$ while it was $V(t)$ at the start and therefore the acceleration, the change of the speed in a time interval Δt, is $a = [V(T + \Delta t) - V(t)]/\Delta t = \Delta V/\Delta t$ if you take the intervals to be small enough. For a constant acceleration, or a constant speed, we can take the interval of time to be as small as you want, as long as the difference in the position and the velocity decrease at the same rate as Δt decreases. This is the basic idea of a *derivative* or, as Newton would call it, the *fluxion*. If the speed, or the space, changes continuously (and that notion is a tricky one that would exercise mathematicians for nearly two centuries), we have a device for calculating the rate of change. For a curve in, for example, two dimensions, (x, y), the curve is $y(x)$ and the rate of change of one coordinate relative to another is $\Delta y/\Delta x$. The continuity is assured when we can take the limit as $\Delta x \to 0$ and have the interval in y change without becoming infinite because $\Delta y \to 0$ at the same rate. Newton used a dot to denote this quantity when it referred to space and time; a more general way of writing this is:

$$V(t) = \lim_{\Delta t \to 0} \frac{\Delta x}{\Delta t} \equiv \frac{dx(t)}{dt} \equiv \dot{x}(t)$$

and similarly for acceleration. This does *not* mean that the speed, or acceleration, is necessarily constant everywhere and forever, merely that it remains constant over an infinitesimal interval, which I hope seems intuitive.

If we imagine that a quantity depends on many things, for instance three different directions in space as well as time, we can ask how it changes when only one of these is changed at a time. For instance, the area of an object depends on two coordinates, the volume on three, etc. So we can imagine a generalized something, call it now a function $F(x, y, z)$, for instance, and then the change of F with respect to only one of the quantities on which it depends, say x, is

$$\lim_{\Delta x \to 0} \frac{F(x + \Delta x, y, z) - F(x, y, z)}{\Delta x} \equiv \frac{\partial F}{\partial x}$$

the other two remaining fixed. This is the *partial derivative* and it is going to be an essential feature of all subsequent discussions of how fields change. Along with the changes, we have the cumulant, the path itself. This is a different problem, to find a way to sum the tiny intervals covered in each interval of time to produce a finite distance. This is the *integral*, first introduced by Leibniz,

$$L = \lim_{N \to \infty} \sum_{k=0}^{N} V(t_k)\Delta t_k \equiv \int_0^T V(t)dt.$$

The sum is then the result of infinitely many infinitesimal intervals during which the speed remains constant, but allowing that the speed may be continually changing from one tiny interval to the next. It is a modern notion that the integral is the inverse operation to derivation, but you can see how it happens. The term "quadrature" refers to taking the integral, while for a geometric discussion when length and not time are involved, the derivative is the tangent to the curve. The idea of a variation is closely related to a differential, the unspecified change of a continuous quantity δt, for instance, that can be made as small as we'd like but is otherwise unconstrained. This was already implicit in statics, from Archimedes and Leonardo. Remember that the lever requires the concept of a virtual displacement, where we imagine taking weights and lifting it at one end while lowering it at the other by some angle $\delta\theta$. What we're now doing is making this a variable

Motion of individual particles can be extended to the behavior of continuous media with the extension of force and so the mathematical machinery also needs to be modified. Now, instead of a trajectory—a line—in space marking the motion in time, we can have simultaneous variations of many points on a surface or a volume at the same time, all parts of which are connected to each other. This is the origin of partial differential equations. The variation of a distributed quantity, say something that depends on two coordinates (x, y), we can take the change in each direction separately keeping the alternative position fixed (that is, the change of the function at one position with respect to the other):

$$\Delta f(x, y) = \left[\frac{\Delta f}{\Delta x}\right]_y \Delta x + \left[\frac{\Delta f}{\Delta y}\right]_x \Delta y \rightarrow \left(\frac{\partial f}{\partial x}\right)_y \Delta x + \left(\frac{\partial f}{\partial y}\right)_x \Delta y.$$

This can be extended to higher order derivatives as well distributed in space. The close link between differential equations and dynamics had already been realized by Euler and Lagrange in the eighteenth century, but was exploited far more in the next. The connection is clear from the second law of motion, that the rate of change of a velocity with time—the acceleration, that is the second derivative of the displacement—is produced by a force that can change in space. It was, therefore, a short step to associate *any* second-order differential equation with forces.[1] The important idea is that a line, surface, or volume can change continually but not identically in both space and time. It depends on the initial and boundary conditions. Much of the theory related to these equations in the nineteenth century involved studying which conditions lead to which types of solutions, whether these are uniquely specified by the boundary conditions, and how they evolve. Fourier was the first to show that the solutions to these equations can be expressed as a series of superimposed waves, periodic functions in the independent coordinates. For example, for a string this is an infinite series of sines and cosines that are the harmonic functions of *modes* of the string (the harmonics of the fundamental frequency). He showed how to use the boundary conditions to obtain the amplitudes—thus determining the phases—of the waves. The method is called the Fourier integral or, more frequently, the Fourier transform. It was a complementary representation to one used by Laplace to solve field problems for gravity, the Laplace transform.

VECTORS AND TENSORS

As we discussed with medieval kinematics, the problem of describing how a body moves was especially vexing when it required more than one dimension. The difference between speed and velocity is that the latter involves a direction. The quantity is called a *vector*, a term coined by William Rowan Hamilton in the mid-1800s long after the thing itself was first employed. But how can you quantify this "object"? The components of a force, or a velocity, are, as Newton showed in the *Principia*, combined using a parallelogram. The resultant of two orthogonal components lies along the diagonal of the resulting figure and the Pythagorean theorem gives its magnitude. Inversely, for any coordinate system to which the components are referred, the single vector is decomposable into its components each of which points along some axis that is represented by a unit vector, $\hat{\mathbf{n}}$ for instance for some normal vector, that has a magnitude of unity and signifies the direction. For example, a wind speed may be V km hr^{-1} from the northeast, but this can also be described as a vector $\mathbf{V} = V_N\hat{\mathbf{N}} + V_E\hat{\mathbf{E}}$ where V_E and V_N are the components. This is a different sort of addition. The two directions, north and east, remain distinct. Hamilton's 1842 realization was the analogy with complex numbers, for which the real and imaginary components remain distinct and that sum quadratically in magnitude, and how to also multiply these quantities. There are two separate forms for this. One is the projection of one vector onto another, the result then being only a magnitude and thus called a *scalar*. Represented for two vectors **A** and **B** by $C = \mathbf{A} \cdot \mathbf{B}$ it doesn't matter which vector is projected onto which. On the other hand, recall our many discussions of the lever. A beam lies along some direction and a force applied perpendicular to its length produces a rotation about an axis that is perpendicular to the two of them. The *torque*, the force that produces the angular rotation, is given by the *vector product*, $\mathbf{T} = \mathbf{r} \times \mathbf{F}$ where **F** is the force on the beam; the angular momentum is $\mathbf{r} \times \mathbf{v}$ for a transverse velocity **v**.[2] An important, indeed fundamental property of this product is that it doesn't commute. If you apply the produce the other way round it reverses sign (the rotation goes in the opposite sense). This decomposition also applies to *changes*. A gradient is the directional change in a quantity and is, therefore, a vector. But a vector quantity can also change in each direction and this requires a further generalization to a *tensor*, another Hamiltonian. More general than vectors, since these describe planes as well as lines in space (for instance, you can say "in a direction" or "lying in the plane"), the foundations were laid somewhat later in the 1890s through the first decade of the twentieth century by Gregorio Ricci-Curbastro and Tullio Levi-Civita under the name "absolute differential calculus." An earlier form, the *dyadic*, was introduced by Josiah Willard Gibbs to describe strains, stresses, and fields. A simple way of imagining the thing is to mark a bunch of points on a sheet of plastic wrap and then stretch the surface along some arbitrary direction. The points displace differentially in the two directions (or three if the surface buckles). You can see how this would apply to the generalization of force from Newtonian to relativistic gravitation, and also how this generalizes the application of a force to a deformable body. A tensor has multiple components as does a vector but these can mix depending on their sense of displacement. So something pointing in the NE direction can have components that both change along the N and E directions, for example. So, for instance, a tensor $\mathcal{T}$ can be written with its components along, say, two directions simultaneously,

$$\mathcal{T} = T_{NN}\hat{\mathbf{N}}\hat{\mathbf{N}} + T_{NE}\hat{\mathbf{N}}\hat{\mathbf{E}} + T_{EN}\hat{\mathbf{E}}\hat{\mathbf{N}} + T_{EE}\hat{\mathbf{E}}\hat{\mathbf{E}} + \cdots.$$

The individual components do not have to be symmetric (e.g., T_{EN} doesn't necessarily have to be the same as T_{NE}. Think of a vortex shearing that has a sense of helicity and you can see why this might be so).

VARIATIONAL CALCULUS

For Fermat, passing light through a refracting medium was a minimization problem, the problem was to find the path that minimized the transit time or one of *minimum action* of the medium. If the speed of light in a vacuum is a constant but the speed within anything else is lower, a continuously changing index of refraction is the same as a continuously varying speed of propagation. Then the path of minimum length may be other than a straight line (which would be the result for a uniform medium). As an interesting combination of differential and integral methods, this question can be addressed without any recourse to dynamics. The length of any segment of the line is ds, say, but the total length is $S = \int ds$ between any two fixed boundary points. The path can be two- or three-dimensional. The minimum length is the same as the minimum time and this is the connection with dynamics. If we ask, instead, what will be the minimum *time* to move between two points under the action of a force, we have the *brachistochrone* problem, which was posed as a challenge to the continental mathematicians by Newton and the Royal Society soon after the appearance of the *Principia*. The problem is essentially the same, if the force is a function of space, to find the path corresponding to the total acceleration experienced by a particle. Then, for a statics problem, the two approaches can be combined to find the curve of minimum length hanging under the action of a constant external force (the *catenary* curve).

In the second half of the eighteenth century, this problem was generalized following Leibniz's ideas of *vis viva* and *action*. Lagrange realized that the principle of least action could be geometricized along the same lines as the brachistochrone, to find the path that minimizes the action required for motion between two fixed points in space and/or time. I'll go into a bit more detail here just to show you how such a calculation looks and how the formalism expressed the physical ideas. Taking a (potential) field that doesn't explicitly depend on time and is independent of the velocity of the particle, the force being the gradient of the potential, the acceleration depends on the change of the momentum in time and is given by a change in the potential in space. He realized he could write the generalized form for the momentum:

$$p = \frac{\partial T}{\partial \dot{x}},$$

which is read as "the momentum along some direction, say x, is the change in kinetic energy (*vis viva*) with respect to the velocity in the x direction," and combine this with the equation for the force:

$$\frac{d}{dt}\left(\frac{\partial T}{\partial \dot{x}}\right) = -\frac{\partial V}{\partial x},$$

which reads "the force is the rate of spatial change of the potential (field) along the same direction," to obtain a new, universal function:

$$L = T - V$$

now called the *Langrangian* function, from which the equations of motion could be rewritten in terms of this single function:

$$\frac{d}{dt}\left(\frac{\partial L}{\partial \dot{x}}\right) = \frac{\partial L}{\partial x}$$

for *each* component of the system of coordinates. But more important was his realization that this equation is a result of the same minimization principle that takes the action and asks what trajectory produces a minimum value for

$$S = \int_{t_0}^{t_1} L dt$$

during an interval of time $\Delta t = t_1 - t_0$, that is,

$$\delta S = 0$$

expresses the least action principle in compact form. The energy need not be constant for this path, and because the function is a scalar it is independent of the coordinate system (its form changes but the two components, potential and kinetic energies, remain distinct and it permits interactions between the particles). This has become the cornerstone of modern physics. Notice the difference between the momentum and velocity, the latter is the "primitive" variable while the former is derived from the derivatives, which generalizes the concept of the particle momentum by allowing even massless particles (e.g., light in the sense of photons) to have a momentum and, in the relativistic framework, allows us to work in any coordinate frame. Of course, we must confirm that this extremum, δS, is really the *minimum* action, which can only be verified by showing that the second variation is negative, $\delta^2 S < 0$. Weirstrauss and the Berlin school of mathematical analysts made this a central feature of their studies by generalizing the Lagrangian to functions in general, founding a whole branch of mathematics called *variational methods* that is no longer directly connected with the dynamical origins of the problem.[3]

The major change in perspective was Hamilton's introduction of a canonical representation for the momentum and position using the energy (the Hamiltonian function) instead of the action as the fundamental link. Although using the least action principle, Lagrange had introduced the idea of a derivative function as the momentum, Hamilton fully exploited its possibilities by associating two ordinary differential equations (in vector notation):

$$\frac{d\mathbf{p}}{dt} = -\frac{\partial H}{\partial \mathbf{q}}$$

for the momentum evolution and

$$\frac{d\mathbf{q}}{dt} = \frac{\partial H}{\partial \mathbf{p}}$$

for the generalized velocity. For Hamilton the position and momentum, *not* the velocity, were the primitive variables (unlike Lagrange and Euler). Notice I've called the first the momentum "evolution," not the force. This is the important change. For any particle, each point in space has an associated momentum and both are needed to specify a trajectory (ultimately $\mathbf{q}(t)$). Then these can become the characteristics, analogous to those Monge had introduced to describe surfaces, along which the particle moves. If the total energy is conserved, the set of coordinates describes fully its history. Actually this is not that different from Lagrange's original conception of the function that bears his name except the Lagrangian is not a conserved quantity while, if H is independent of time, the Hamiltonian function is. Since these together, for a closed path (an orbit), require only the phase instead of the time, the space they collectively describe is called a *phase space* and this was taken over directly into statistical mechanics by Boltzmann and Gibbs.

GROUPS

Although a sidelight in our development, the nineteenth century also witnessed an explosive growth in mathematical physics, unprecedented in the previous two centuries. New viewpoints emerged, linking a broad range of mathematical ideas. One of the most fruitful for the concept of force, and dynamics, was the study of transformation groups that formed the nucleus of Sophus Lie's (1842–1899) contributions to mathematics. We can develop this rather simply from the Hamiltonian perspective. In that approach, a particle's trajectory is the evolution of its momentum and position through time. These are linear differential equations that depend only on the potential.

Lie's theory begins with the linear transformation, one that takes the value of a function at some "place," x, and asks what the value is at $x + \delta$. This is just the definition of the first derivative, the rate of change of the function is df/dx. But we can look at this as well in terms of an operation, one that takes $f(x)$ and returns $f(x + \delta)$. For a very small displacement in the dependent variable, this is developed only to first order in which case, if we say there is a transformation T_δ then $T_\delta f(x) = f(x + \delta)$ and some second $T_{\delta'}$ that is applied to this second function takes it to δ' so that $T_{\delta+\delta'} = T_{\delta'} T_\delta$. The inverse of this transformation also exists, the path is reversible, so there is also an operator T_δ^{-1} for every T_δ. The product of these two is the identity operator, the function is left unchanged. Now if we change our language and think in terms of motion, you'll see this is another way to look at the dynamics. It's easy to see that if we've chosen a line, the transformations are also commutative, it doesn't matter in which order we take them and the same is true for their inverses, and any other combination. For a surface, however, we don't necessarily get the same result. It depends what we mean by a transformation. If we've moved the particle on a two-dimensional path, the velocity vector is always tangent to the curve. Let the surface be a plane and the paths be orthogonal displacements on two arbitrary paths. Although we may arrive at the same point in the end, the velocity is not necessarily in the same direction. Now if we add the restriction the curve is closed; displacing around a closed curve (an orbit) on this plane leaves the vector pointing in the same direction as at the start. Now let's do this on a curved surface. Then we get a rotation of the vector, as Lie's idea was to replace the specific derivative operator with one that is more general, a "displacement." His treatment followed on the study by Plücker in the 1860s and Riemann at around the same time, realizing that translations (in space and time) have an interesting mathematical structure.

Dynamical symmetry arguments were also used by Einstein in his 1905 introduction of relativity, what has since become known as the Lorentz group. This set of transformations in spacetime satisfies the group properties of invertability, association, and also commutivity. It sits at the core of the theory, the constant speed of light providing the means for going between reference frames. If instead of writing the operation out explicitly we use Λ to represent the transformation and γ to be the governing parameter (this is the factor $\gamma = 1/\sqrt{(1 - v^2/c^2)}$ in the Lorentz transformations), we have the property that for two frames in relative motion $u = (v_1 + v_2)/(1 + v_1 v_2/c^2)$ for the velocity addition law and for successive transformations $\Lambda = \Lambda_2 \Lambda_1$ if we go from one frame to another. In another way, simpler static, geometric symmetries are obvious in crystals and during the early modern period these were likened to the Platonic solids, thus to ideal forms. Kepler used this in his *Mysterium Cosmologica* to imbed the orbits of the planets in the five solids and wrote treatises on the symmetry of snowflakes and the dodecahedron. This symmetry of crystals is a basic feature of quantum mechanics applied to the solid state. It is based on a theorem due to Felix Bloch, that a free electron is sort of confused when being propagated in a periodic structure and can be expanded in an elementary set of periodic functions on the scale of the lattice. The same is used for computing the vibrational spectrum of crystals, railroad bridges, and other composite structures that possess specific symmetries.

DIFFERENTIAL GEOMETRY: ILLUSTRATED WITH THERMODYNAMICS

Producing maps was a preoccupation, indeed a mania, for the imperialist expansionists of the eighteenth and nineteenth century. Surveying was a highly developed technical art. But the maps were another matter. Coordination of measurements on a complex terrain is hardly as simple a problem as measuring the altitude and bearing of a set of points on the surface. In particular, projective geometry developed to respond to this immediate problem and then was generalized when the mathematical problem appeared far more general than the specific applications. For instance, on a sphere, how do we take two regions measured on what looks like a flat plane and then distort them appropriately and match their boundaries to form a continuous surface? As you saw in the description of relativity, this later, through the ideas of parallel transport and curvature, became fundamental tools for studying trajectories and led to the tensor formalism of general relativity.

We can use a remarkable feature of thermodynamics, its connection with geometry in an unexpected way, to illustrate how differential geometry is applied to a physical problem. Small thermal changes, calorimetry, were described in the mid-nineteenth century using linear equations and this could easily be adapted to the statement of the first law, that a quantity of heat added to a system produces both changes in the internal state of the system and work. Now since the internal energy depends only on the temperature, a cyclic change should leave it at the same value, just as a path in a gravitational field brings us back to the same place. For that reason it is called a *potential*. In contrast, the work depends on two state variables, pressure and volume, and around a closed path the total work is due to a change in the heat content, not just the internal energy. These can be thought of as coordinates, as in an indicator diagram (see the discussion of Carnot's cycle), a reference system in which surfaces represent free energies and entropy. Small changes are separable, and as long as the processes are reversible we can use the continuity conditions available for any continuous surface, exactly as in geometry. This geometricization, surprisingly, furnishes a closer link to dynamics than the usual conceptions and was employed by both Gibbs and Maxwell when describing energies. Gibbs actually sent Maxwell a model surface for the thermal properties of water, which shows not only the complex shape of the functions needed to describe the processes but also the trajectories followed by specific processes. When we described a change in heat, ΔQ, as something that produces work, we can recall that if P is the pressure, V the volume, T the temperature, and U the internal energy,

$$\Delta Q = dU - dW = AdT + BdV = dU(T) + PdV$$

so the independent coordinates are (T, V) and the length of the infinitesimal displacements become the unit vectors. You'll notice the close correspondence with Lie's ideas of operators. Continuity says when we look at the rates of change relative to the two coordinates, it doesn't matter in which order the changes are taken. Thus, in the language we described for partial derivatives,

$$\left[\frac{\partial A}{\partial V}\right]_T = \left[\frac{\partial B}{\partial T}\right]_V$$

where the subscript tells which quantity is kept constant during the change. Now from the second law of thermodynamics we have a new way of writing the heat, $\Delta Q = T\,dS$ where now S is the entropy, so a surface of constant entropy is one for which we move along a trajectory driven only by T and V since the pressure, P, is a function of only these two. You'll notice this has the same features as the previous discussion of vectors and tensors. Green's insights on potentials again allow us to see that dS is the normal to the entropy "surface" and adiabatic surfaces are the same as equipotentials. We can then form a variety of such thermodynamic quantities, the free energy, $F = U + PV - TS$, the enthalpy, $H = U + PV$, and the Gibbs free energy, $G = F - \mu N$, using the variable μ to represent the chemical potential (the energy required to change the number of particles in the system). In each case, we have a different set of state variables and can transform between them precisely as we would between coordinates in different reference systems. The way we've written these should look now like vectors and tensors. I chose this example precisely because it's so comparatively distant from your usual notions of changes in space and time and to show the conceptual basis of any field theory. But these are not merely formal analogies. Any change can be visualized within a space, this was emphasized by Poincaré at the end of the nineteenth century. The modern terminology for chemical reactions is *landscape*, which use both the quantum mechanical energies and the thermodynamic concepts to describe the favorable states. The rest of the connection between dynamics and differential geometry comes from saying that a thermal system follows a path of least action and the equilibrium states are given by the minima on the thermodynamic potential surfaces. Such a construction was attempted by C. Catheodory in the 1920s and led to an axiomatization of the mathematical structure.

NOTES

1. This would later prove a powerful analytic tool guiding the construction of analog computers that, by judicious couplings of pulleys and springs, later gears, and finally vacuum tubes, would allow the solution of specific individual and systems of equations. In this way, the concept of force was generalized to include a purely analogical role that nevertheless proved essential.

2. The modern notation was introduced by Gibbs at the end of the nineteenth century when vector analysis replaced quaternions. This is also written as a "wedge product," $\tau = \mathbf{r} \wedge \mathbf{L}$, depending on the cultural habits of the authors. The notation was introduced by Hermann Grassmann at around the same time as Hamilton in his analysis of forms.

3. As a more detailed illustration, for those of you who have seen some calculus, I'll treat a simple but fundamental problem: how to use the variational method to solve a lever or Atwood machine problem. The principle of least action, and the extension of the Atwood machine, unites angular momentum and the lever in a way that extends Archimedes' reasoning and generalizes the result. We consider two masses m_1 and m_2 on opposite sides of the fulcrum of a lever at distances r_1 and r_2. The most important feature of this simple picture is that the two masses are constrained to move, as in the Atwood machine, along some path governed by a common property. In the latter case, the length of the cord connecting the masses doesn't change and this is the same example used by both D'Alembert and Lagrange in the block and tackle couplings with which they illustrated virtual displacement. For a level it's more obvious. The two masses move through a common *angle*, ϕ, provided the lever remains rigid. The vertical distances moved are, however, different because their distances from the pivot point are not equal. For a simple Archimedian example, the masses are not dynamical and the equilibrium is obtained by forcing the equality

$$m_1 r_1 g = m_2 r_2 g \leftrightarrow W_1 r_1 = W_2 r_2.$$

The transformation from force to potential energy makes clearer the Greek mechanician's fundamental axiom. Note that this is true for an arbitrary displacement no matter how small. The extension comes when the least action principle is applied to the dynamical system. The kinetic energy is quadratic in the angular displacement so

$$T = \frac{1}{2}\left[m_1 r_1^2 + m_2 r_2^2\right](\dot{\phi})^2$$

where $\dot{\phi} = \omega$, the angular velocity (or frequency), again for any arbitrary angular displacement and

$$P = (-m_2 r_1 + m_2 r_2) g \phi$$

for the potential energy. The signs of the two masses, although they must be opposite, are arbitrary. Notice that imposing the constraint is that *the angular displacements are the same*, that is we have a *holonomic system*; we arrive at

$$\left(m_1 r_2^2 + m_2 r_2^2\right)\ddot{\phi} \equiv I\alpha = -(W_1 r_1 + W_2 r_2)\phi$$

which is the equation of motion for a barbell with an angular acceleration α. It was then simple to extend this to any arbitrary number of masses extending along the rod with the subsequent *definition* of the moment of inertia, I, as

$$I = \sum_i m_i r_r^2 \rightarrow \int r^2 dm,$$

where the integral is the continuous limit of the mass distribution. This also yields the radius of gyration since the system is rigid and this can be defined independent of time. The formalism thus connects immediately to the motion of a pendulum for which the length is only r_1 and with only a single mass and yields a harmonic oscillator with the usual well-known frequency. Although hardly still necessary at the end of the eighteenth century, this also clearly resolved the question of the equilibrium of, and oscillation of, a bent beam since it only requires a change in the angles for the calculation of the moment of inertia.

TIMELINE

5th–4th century BC	Plato and the school of Athens; Eudoxus of Cnidus, first geometric-kinematic model of planetary motion; Euclid *Elements*, compilation of hellenistic geometry; Aristotle (*Physica*, *de Caelo*), laws of motion and nature of space and time; Democritus and Epicurus of Samos and atomism
3rd century BC	Archimedes, hydrostatics, mechanical principles of equilibrium; Apollonius of Perga, kinematical theory of epicyclic motion for planetary astronomy; Aristarchus of Samos, first suggestion of a heliocentric arrangement of the planets
2nd century BC	Hipparchus and precession
1st century BC	Lucretius, *de Rerum Natura*, atomistic philosophy
1st century AD	Vitruvius; Hero of Alexandria
2nd century AD	Claudius Ptolemy, *Almagest*, compilation of hellenistic mathematical planetary astronomy
800–900	Thabit ibn Qurra's *Kitab fi'l-qarastun* (The book on the beam balance)
1000–1037	ibn Sina (Avicenna)
1100–1200	ibn Rushd (Averroes); John of Seville translates *Almagest*; Gerard of Cremona active in translation; Maimonedes
1200–1237	Jordanus de Nemore, positional gravity and the origin of medieval statics
1210, 1277	Two principal condemnations of non-Aristotelian philosophy issued at Paris
1268–1286	William or Moerbeke's translations of Greek works on mathematics and mechanics

1320–1450	Thomas Bradwardine and the Merton school of "calculators," laws of motion, the mean value theorem ("the Merton Rule") for displacement and speed in time; Jean Buridan, Nichole Oresme and the Parisian school of mechanics, introduction of graphical methods, studies of the laws for continuous and accelerated motions; Nicholas of Cusa, problem of relative motion of the observer and motion of the earth.
1470–1505	Leonardo da Vinci and applied mechanics
1472	von Peuerbach's *Theoriae Novae Planetarum.*
1492	Columbus' first voyage to the Americas
1542	Copernicus' *de Revolutionibus Orbium Coelestium*, introduction of a consistent heliocentric planetary model
1564	birth of Galileo
1588	Tycho's *De mundi aetherei recentioribus phaenomenis*, publication on the comet of 1577 and the introduction of a composite helio/geocentric cosmology
1600	Gilbert's *de Magnete*
1609	Kepler's *Astronomia Nova*
1610	Galileo's *Sidereus Nuncius*
1618–1621	Publication of Kepler's *Epitome astronomiae Copernicanae*; Galileo's *il Saggiatore*
1632	Galileo's *Dialogo sopra i due massimi sistemi del mondo*
1638	Galileo's *Discorsi e dimostrazioni matematiche intorno a due nuove scienze*
1642	Death of Galileo, birth of Newton
1644	Descartes' *Principia Philosophiae*; Torricelli's announcement of the invention of the barometer
1647	Pascal's demonstration of atmospheric pressure gradient
1658	Christian Huygens' description of the pendulum clock and centrifugal force
1662	Boyle's law for gas pressure and studies of vacuum and gases
1665–1666	Newton discovers the inverse square law for gravitation (the *annus mirabilis*)
1672	von Guericke's demonstration at Magdeburg of atmospheric pressure
1676–1689	Leibniz' dynamical investigations, especially principle of *vis viva*
1679	Hooke's law presented in the *Cutlerian Lectures*
1687	Newton's *Principia Mathematicae Naturalis*; 2nd edition 1713; 3rd edition 1726; first English translation (Andrew Motte) 1729
1695–1697	Halley's papers on the tides
1704	First edition of Newton's *Optiks*
1712	Newcomen invents the modern steam engine
1725	Discovery of abberation of starlight by Bradley

1727	Death of Newton
1734	Voltaire's *English Letters* (*Lettres philosophiques*) that popularized the Newtonian philosophy on the continent
1736–1737	Euler's *Mechanica*
1738	Publication of Daniel Bernoulli's *Hydrodynamica*, derivation of Boyle's law, and first comprehensive treatise on fluid dynamics
1743	D'Alembert's *Traité de dynamique* and the principle of virtual work
1747–1760	Franklin's experiments and studies on electricity
1748	Maclaurin's *Account of Sir Issac Newton's Philosophical Discoveries*
1750–1780	Debate on the nature of charge
1754–1756	Lagrange and Euler's foundation of calculus of variations
1758	Return of the comet of 1682, now called "Halley's comet'; publication of Boscovich's *Theoria Philosophae Naturalis*'
1759	Aepinus publishes the two-charge theory of electricity
1769	Watt improves the Newcomen design with a separate condenser
1775	Maskylene's measurements of the density of mountains, foundations of geodesy
1781	William Herschel discovers Uranus
1783	Death of Euler
1784	Coulomb torsion balance experiment on electric charge; Atwood's description of the simplest mechanical device
1788	Lagrange's *Traité de mécanique analytique*
1789	Lavoisier's *Elements of Chemistry*
1798	Cavendish experiment, determination of the density of the Earth (and the Newtonian gravitational constant)
1799	Beginning of the publication of Laplace's *Mecanique Celeste*
1813	Gauss' law for fields; publication by Poisson of the field equation for a matter/charge distribution
1820	Oersted's discovery of electromagnetism
1821	Publication by Biot and Savart of the law for electromagnetism
1821–1823	Amprere's publication of the laws of electrodynamics
1824	S. Carnot's *On the Motive Power of Fire*
1827	Green's publication of *Essay on Electricity and Magnetism*; Ohm's explanation for circuits and currents
1828	Brown's paper on microscopic motion of pollen (Brownian motion)
1831	Faraday and Henry discover electromagnetic induction and self-induction
1831–1851	Faraday's *Experimental Researches in Electricity and Magnetism*

1834–1835	Hamilton's *On a General Method in Dynamics*
1842	Hamilton's discovery of quaternions and vector analysis
1843	Joule's determination of the mechanical equivalent of heat
1846	Discovery of Neptune, predicted by Adams and Leverrier based on Newtonian gravitational theory
1847	Publication of Helmholtz's *The Conservation of Forces*
1851	Foucault's demonstration of the Coriolis effect and rotation of the Earth
1852	Invention of gyroscope; Faraday publishes the idea of "lines of force"
1857	Maxwell publishes *On Faraday's Lines of Force*; publication of Clausius' *A Kind of Motion We Call Heat*; Kelvin and Helmholtz determine the lower limit to the age of the Sun
1860	Maxwell's *On the Dynamical Theory of Gases*
1862	Maxwell's *On the Dynamical Theory of the Electromagnetic Field*
1867	Thompson and Tait's *Principles of Natural Philosophy*
1871–1906	Boltzmann's work on statistical mechanics, *Lectures on Gas Theory*
1872	Maxwell's *Treatise on Electricity and Magnetism*
1881–1887	Michelson, along with Morley, determines invariance of the speed of light
1883–1901	Mach's *Principles of Mechanics* (four editions)
1884	Poynting's paper on radiation pressure
1887–1890	Poincaré on three-body problem and orbital stability, chaos in dynamical systems
1892	Lorentz's paper on the electrodynamics of moving bodies
1898	J. J. Thomson's determination of the mass and charge of the electron
1900	Planck introduces quantized statistics for blackbody radiation
1902	Gibbs publishes *Elementary Principles of Statistical Mechanics*
1905	Einstein's papers on Brownian motion, special relativity theory, and first paper on the photoelectric effect
1907	Minkowski introduces spacetime, the unification of space and time
1909	Perrin measures molecular trajectories using Einstein's theory of Brownian motion; Milliken and Fletcher's oil drop experiment for the elementary unit of charge
1911	Rutherford's scattering experiments; Einstein derives the precession rate of Mercury's orbit
1913	Bohr's papers on atomic structure and the Correspondence Principle, *On the Constitution of Atoms and*

	Molecules; Einstein and Grossmann's *Entwurf* paper on generalized gravitation theory
1916	Einstein publishes the final General Theory of Relativity; Schwarzschild obtains first solution for a nonrotating point mass
1919	Measurement of deflection of light during a solar eclipse
1919–1923	Sommerfeld's *Atomic Structure and Spectral Lines*
1922	Compton's discovery of dynamical X-ray scattering, Compton effect
1923-1924	de Broglie's thesis on wave-particle duality
1925	Heisenberg's paper on quantum mechanics
1925–1926	Born, Heisenberg, and Jordan's papers on quantum mechanics; Heisenberg's uncertainty principle
1926–1927	Schrödinger's foundational papers on wave mechanics
1927	Bohr introduces complementarity
1928	Dirac's equation for relativistic electrons and the beginning of quantum electrodynamics
1930	Dirac's *Principles of Quantum Mechanics* (3rd edition 1965); Born and Jordan's *Elementare Quantenmechanik*
1932	Discovery of the neutron
1934	Fermi's theory of β-decay and the weak interaction
1934–1970	Ingredients of the "Standard Model" for electroweak unification
1935	Yukawa's meson theory of nuclear structure
1948	Feynman's paper introducing the path integral formalism for quantum mechanics
1974	Discovery of the binary pulsar 1915+21 and gravitational wave emission
1979	Discovery of cosmological scale gravitational lenses

BIBLIOGRAPHY

Agazzi, J. 1971. *Faraday as a Natural Philosopher* (Chicago).

Aiton, E. J. 1972. *The Vortex Theory of Planetary Motions* (New York: Neale Watson).

———. 1987. "Peurbach's *Theoricae Novae Planetarum*: A Translation with Commentary." *Osiris*, **3**, 4.

Aristotle. 1930. *Physica, de Caelo* (trans. W. D. Ross) (Oxford: Oxford University Press).

———. 1936, *Minor Works: Mechanics* (trans. W. S. Hett) (Cambridge, MA: Harvard University Press)

Atwood, G. 1794. "Investigations, Founded on the Theory of Motion, for Determining the Times of Vibrations of Watch Balances." *Phil Trans R. Soc.*, **84**, 119.

Blackwell, R. J. (ed.) 1986. *Christian Huygens' "The Penulum Clock or Geometrical Demonstration Concerning the Motion of Pendula as Applied to Clocks"* (Ames: Iowa State University Press).

Boltzmann, L. 1995. *Lectures on Gas Theory* (ed. S. G. Brush) (New York: Dover).

Boscovich, S. J. and Roger, J. 1763. *A Theory of Natural Philosophy* (trans. J. M. Childs) (Chicago: Open Court) (ed. 1922).

Brush, S. G. 1976. *The Kind of Motion We Call Heat: A History of the Kinetic Theory of Gases in the 19th Century* (2 vols.) (New York: American Elsevier).

Brush, S. G. and Hall, N. S. 2003. *The Kinetic Theory of Gases: An Anthology of Classic Papers with Historical Commentary* (London: Imperial College Press).

Carmody, T. and Kobus, H. (eds.) 1968. *Hydrodynamics, by Daniel Bernoulli, and Hydraulics, by Johann Bernoulli* (New York: Dover).

Casper, M. 1959. *Johannes Kepler* (New York: Dover).

Chandrasekhar, S. 1967. *Ellipsoidal Figures in Equilibrium* (New York: Dover).

———. 1995. *Newton for the Common Reader* (Cambridge, UK: Cambridge University Press).

Clagett, M. 1959. *The Science of Mechanics in the Middle Ages* (Madison: University of Wisconsin Press).

———. 1964. *Archimedes in the Middle Ages* (multiple volumes) (Wisconsin: University of Wisconsin Press) (Philadelphia, PA: American Philosophical Society).

Cohen, I. B. 1980. *The Newtonian Revolution* (Cambridge, MA: Harvard University Press)

———. 1985. *Revolutions in Science* (Cambridge, MA: Harvard University Press).

———. 1991. *The Birth of the New Physics* (2nd ed.) (New York: W. W. Norton).

Cohen, I. B. and Drabkin, I. E. 1948. *A Source Book in Greek Science* (Cambridge, MA: Harvard University Press).

Cohen, I. B. and Whitman, A. (eds.) 1999. *Isaac Newton—The Principia: Mathematical Principles of Natural Philosophy* (Berkeley: University of California Press) (also see Cohen, I. B. and Koyrè, A. 1972, *Isaac Newton's Philosophiae Naturalis Principia Mathematica*, Facsimile of third edition (1726) with variant readings, 2 vols.) (Cambridge, MA: Harvard University Press).

Copernicus. 1938–1941. *de Revolutionibus* Preface and Part of Vol. 1 plus notes. *Occas. Notes R. Astr. Soc.* **1**, 1

Crombie, A. C. 1959. *Medieval and Early Modern Science* (2 vols.) (Garden City, NY: Anchor Books).

———. 1971. *Robert Grossteste and the Founding of Experimental Science* (Oxford: Oxford University Press).

Crossby, H. L. (ed.) 1961. *Thomas of Bradwardine—His Tractatus de Proportionibus* (Madison: University of Wisconsin Press).

Darrigol, O. 2000. *Electrodynamics from Amperè to Einstein* (Oxford: Oxford University Press).

Dijsterhuis, E. 1987. *Archimedes* (Princeton, NJ: Princeton University Press).

Dirac, P. A. M. 1965. *The Principles of Quantum Mechanics* (3rd ed.) (Oxford: Oxford University Press).

———. 1978. *Directions in Physics* (New York: Wiley).

Dobbs, B. J. T. 1975. *The Foundations of Newton's Alchemy or 'The Hunting of the Greene Lyon'* (Cambridge, UK: Cambridge University Press).

Drabkin, I. E. and Drake, S. 1960. *Galileo on Motion* (Madison: University of Wisconsin Press).

Drachmann, A. G. 1963. *The Mechanical Technology of the Greeks and Romans* (Madison: University of Wisconsin Press).

Drake, S. (ed.) 1953. *Galileo Galilei: Dialogue Concerning the Two Chief World Systems* (Berkeley: University of California Press).

Drake, S. 1957. *Discoveries and Opinions of Galileo* (New York: Anchor).

Drake, S. (trans, ed.) 1974. *Galileo: Two New Sciences* (Madison: University of Wisconsin Press).

Drake, S. 1978. *Galileo at Work: His Scientific Biography* (Chicago: University of Chicago Press).

Drake, S. 1981. *Cause, Experiment, and Science: A Galilean Dialogue, Incorporating a New English Translation of Galileo's Bodies That Stay Atop Water, or Move in It Cause, Experiment, and Science* (Chicago: University of Chicago Press).

Drake, S. and Drabkin, I. E. 1969. *Mechanics in Sixteenth-Century Italy: Selections from Tartaglia, Benedetti, Guido Ubaldo, and Galileo* (Madison: University of Wisconsin Press).

Dreyer, J. L. E. 1953. *A History of Astronomy from Thales to Kepler* (New York: Dover).

Effler, R. R. 1962. *John Duns Scotus and the Principle of "Omni Quod Movetur ab alio Movetur"* (New York: Fraciscan Inst. Publ.).

Einstein, A. 1990. *Collected Papers of Albert Einstein* (vols. 2 and 6) (Princeton, NJ: Princeton University Press).

Elkana, Y. 1974. *The Discovery of the Conservation of Energy* (Cambridge, MA: Harvard University Press).

Freeman, K. 1977. *Ancilla to the Pre-Socratic Philosophers* (Cambridge, MA: Harvard University Press).

Gaukroger, S. 1995. *Descartes: An Intellectual Biography* (Oxford: Oxford University Press).

Gibbs, J. W. 1949. *The Collected Papers of J. Willard Gibbs* (vol. 1) (New Haven, CT: Yale University Press).

Gilbert, W. 1958. *On the Magnet: de Magnete, a Translation by the Gilbert Society, 1900* (New York: Basic Books).

Gillespie, H. 1960. *The Edge of Certainty: An Essay in the History of Scientific Ideas* (Princeton, NJ: Princeton University Press).

Gilmore, C. S. 1971. *Coulomb and the Evolution of Physics and Engineering in Eighteenth-Century France* (Princeton, NJ: Princeton University Press).

Goddu, A. 2001. "The Impact of Ockham's Reading of the *Physics* on the Mertonians and Parisian Terminists." *Early Sci. Med.*, **6**, 204.

Goldstein, H. H. 1980. *A History of the Calculus of Variations from the 17th Century through the 19th Century* (Berlin: Springer-Verlag).

Gowing, R. 1983. *Roger Cotes: Natural Philosopher* (Cambridge University Press).

Grant, E. 1977. *Physical Science in the Middle Ages* (New York: Wiley).

———. 1981. *Much Ado About Nothing: Theories of Space and Vacuum from the Middle Ages to the Scientific Revolution* (Cambridge, UK: Cambridge University Press).

———. 1996. *Planets, Stars, and Orbs: The Medieval Cosmos 1200–1687* (Cambridge, UK: Cambridge University Press).

Gratton-Guiness, I. 1995. "Why Did George Green Write His Essay of 1828 on Electricity and Magnetism?" *Amer. Math. Monthly*, **102**, 387.

Green, G. 1871 (rpt. 1970). *The Mathematical Papers of George Green* (New York: Chelsea).

Hall, A. R. 1952. *Ballistics in the Seventeenth Century* (Cambridge, UK: Cambridge University Press).

Hall, A. R. and Hall, M. B. 1962. *Unpublished Scientific Papers of Isaac Newton* (Cambridge, UK: Cambridge University Press).

Hamilton, W. R. 1931. *The Mathematical Papers of William Rowan Hamilton, Vol. 2: Dynamics* and *Vol. 3: Algebra* (Cambridge, UK: Cambridge University Press).

Hardie, R. P., Gaye, R. K., and Stocks, J. L. (trans.) 1930. *The Works of Aristotle: Physica, De Caelo* (Oxford: Oxford University Press).

Harrison, E. 2000. *Cosmology, The Science of the Universe* (2nd ed.) (Cambridge, UK: Cambridge University Press).

Hart, I. B. 1963. *The Mechanical Investigations of Leonardo da Vinci* (Berkeley: University of California Press).

Heath, T. L. 1897. *The Works of Archimedes* (New York: Dover).

Heilbron, J. L. 1979. *Electricity in the 17th and 18th Centuries: A Study of Early Modern Physics* (Berkeley: University of California Press).

Herivel, J. L. 1965. *The Background to Newton's Principia: A Study of Newton's Dynamical Researches in the Years 1664–84* (Oxford: Oxford University Press).

Hertz, H. 1956. *The Principles of Mechanics* (version of 1899) (New York: Dover).

Hesse, M. 1961. *Forces and Fields* (New York: Dover).

Holton, E. 1988. *Thematic Origins of Scientific Thought from Kepler to Einstein* (Cambridge, MA: Harvard University Press).

Jaki, S. L. 1969. *Paradox of Olbers Paradox: A Case History of Scientific Thought* (London: Herder).

Jammer, M. 1999. *Concepts of Force: A Study in the Foundations of Dynamics* (New York: Dover).

Jardine, N. 1982. "The Significance of the Copernican Orbs." *JHA*, **13**, 168.

Joule, J. P. 1850. "On the Mechanical Equivalent of Heat." *Phil Trans. R. Soc.* **140**, 61.

Jourdain, P. E. B. (ed.) 1908. *Abhandlungen über die Prizipiender Mechanik von Labgrange, Rodrigues, Jacobi, und Gauss (Ostwald'sKlassiker der Exacten Wissenschaften Nr. 167* (Leipzig: Wilhelm Engelmann).

———. 1917. *Supplement to Ernst Mach's "The Science of Mechanics 3rd Engl. Ed."* (Chicago: Open Court).

Jungnickel, C. and McCormack, R. 1986. *The Intellectual Mastery of Nature: Theoretical Physics from Ohm to Einstein, Vol. 2: The Now Mighty Theoretical Physics 1970–1925* (Chicago: University of Chicago Press).

———. 1999. *Cavendish: An Experimental Life* (Lewisburg, PA: Bucknell University Press).

Kirschner, S. 2000. "Oresme on Intension and Remission of Qualities in His Commetary on Aristotle's *Physics*." *Vivarium*, **38**, 255.

Koertge, N. 1977. "Galileo and the Problem of Accidents." *J. Hist. Ideas*, **38**, 389.

Koyrè, A. 1955. "A Documentary History of the Problem of Fall from Kepler to Newton." *Trans. Amer. Phil. Soc.*, **45**(4), 329.

———. 1956. *From the Closed World to the Infinite Universe* (Baltimore, MD: Johns Hopkins University Press).

———. 1967. *Newtonian Studies* (Cambridge, MA: Harvard University Press).

———. 1973. *The Astronomical Revolution: Copernicus, Kepler, and Galileo* (Ithaca, NY: Cornell University Press).

Kragh, H. and Carazza, B. 2000. "Classical Behavior of Macroscopic Bodies from Quantum Principles: Early Discussions." *Arch. Hist. Exact Sci.*, **55**, 43.

Kuhn, T. S. 1957. *The Copernican Revolution: Planetary Theory in the Development of Western Thought* (Cambridge, MA: Harvard University Press).

Lanczos, C. 1970. *The Variational Principles of Mechanics* (Toronto: University of Toronto Press).

Laplace, P. S. 1965. *Mecanique Celeste* (trans. N. Bowditch, 5 vols.) (New York: Chelsea).

Lorentz, H. A., Einstein, A., Minkowski, H., and Sommerfeld, A. 1923. *The Principle of Relativity* (New York: Dover).

MacCurdy, E. 1939. *The Notebooks of Leonardo da Vinci* (New York: George Baziller).

Mach, E. 1907. *The Science of Mechanics* (7th ed.) (Chicago: Open Court).

Maier, A. 1982. *On the Threshold of Exact Science: Selected Writings of Anneliese Maier on Late Medieval Natural Philosophy* (ed. S. D. Sargent) (Philadelphia: University of Pennsylvania Press).

Maxwell, J. Clerk. 1920. *Matter and Motion* (Cambridge, UK: Cambridge University Press; rpt. 1991, New York: Dover).

———. 1965. *The Scientific Papers of James Clerk Maxwell* (ed. E. Nevins) (Rpt., New York: Dover).

McMullin, E. (ed.) 1968. *Galileo: Man of Science* (New York: Basic Books).

Meli, D. B. 1993. *Equivalence and Priority: Newton versus Leibniz* (Oxford: Oxford University Press).

Mendoza, E. (ed.) 1960. *Reflections on the Motive Power of Heat: Sadi Carnot, E. Clapeyron, and R. Clausius* (New York: Dover).

Miller, A. I. 1986. *Frontiers of Physics: 1900–1911, Selected Essays* (Zurich: Birkhäuser).

———. 1998. *Albert Einstein's Special Theory of Relativity: Emergence (1905) and Early Interpretation (1905–1911)* (New York: Springer-Verlag).

Moody, E. A. 1951. "Galileo and Avempace: The Dynamics of the Leaning Tower Experiment: I, II." *J. Hist. Ideas*, **12**, 163, 375.

Moody, E. and Clagett, M. 1952. *The Medieval Science of Weights* (Wisconsin: University of Wisconsin Press).

Needham, J. 1962. *Science and Civilization in China, Vol. IVA: Physics* (Cambridge, UK: Cambridge University Press).

Ottman, J. and Wood, R. 1999. "Walter of Burley: His Life and Works." *Vivarium*, **37**, 1.

Pedersen, O. 1981. "The Origins of the 'Theorica Planetarum.'" *JHA*, **12**, 113.

Pedersen, O. and Pihl, M. 1972. *Early Physics and Astronomy* (New York: Science History Publ.).

Petoski, H. 1994. *Design Paradigms: Case Histories of Error and Judgment in Engineering* (Cambridge, UK: Cambridge University Press).

Piaget, J. and Inhelder, B. 1958. *The Growth of Logical Thinking from Childhood to Adolescence* (New York: Basic Books).

Pines, S. (ed.) 1963. *Moses Maimonides: The Guide of the Perplexed* (2 vols.) (Chicago: University of Chicago Press).

Pugliese, P. J. 1989. "Robert Hooke and the Dynamics of a Curved Path" in M. Hunter and S. Schaffer (eds.), *Robert Hooke: New Studies* (London: Boyden Press), p. 181.

Reti, L. (ed.) 1974. *The Unknown Leonardo* (New York: McGraw-Hill).

Rouse, H. and Ince, S. 1957. *History of Hydraulics* (New York: Dover).

Rumford (Benjamin Count of Rumford) 1789. "An Inquiry Concerning the Source of the Heat Which Is Excited by Friction." *Phil Trans. R. Soc.*, **88**, 80.

Sarton, G. *An Introduction to the History of Science* (5 vols.) (Baltimore, MD: Williams and Wilkins).

Schwinger, J. ed. 1958. *Selected Papers on Quantum Electrodynamics* (New York: Dover).

Shapin, S. 1996. *The Scientific Revolution* (Chicago: University of Chicago Press).

Sommerfeld, A. 1943. *Mechanics* (New York: Academic Press).

Southern, R. 1986. *Robert Grossteste: The Growth of an English Mind in Medieval Europe* (Oxford: Oxford University Press).

Spires, I. H. B., Spures, A. G. H., and Barry, F. (eds.) 1937. *The Physical Treatises of Pascal* (New York: Columbia University Press).

Stephenson, B. 1987. *Kepler's Physical Astronomy* (Princeton, NJ: Princeton University Press).

Thijssen, J. M. M. H. 2004."The Burdian School Reassessed: John Buridan and Albert of Saxony." *Vivarium*, **42**, 18.

Thomson, W. (Lord Kelvin) 1882. *The Mathematical and Physical Papers of Lord Kelvin* (5 vols.) (Cambridge University Press).

Thomson, W. and Tait, P. G. 1912. *Elements of Natural Philosophy* (2 vols.) (New York: Dover).

Thorndike, L. 1923–1958. *A History of Magic and Experimental Science* (8 vols.) (New York: Columbia University Press).

Todhunter, I. 1873 (rpt. 1962). *A History of the Mathematical Theories of Attraction and the Figure of the Earth (from the Time of Newton to That of Laplace)* (New York: Dover).

Tricker, R. A. R. 1965. *Early Electrodynamics: The First Law of Circulation* (Oxford: Pergamon).

von Helmholtz, Hermann. 1971. "The Conservation of Force: A Physical Memoire (1847)," in R. Kahl (ed.), *Selected Writings of Hermann von Helmholtz* (Middletown, CT: Wesleyan University Press), p. 2.

Wallace, W. A. 1978. "Causes and Forces in Sixteenth-Century Physics." *Isis*, **69**, 400.

Weisheipl, J. A. 1959. "The Place of John Dumbleton in the Merton School." *Isis*, **50**, 439.

———. 1963. "The Principle Omnequod Movetur ab alio movetur in Medieval Physics." *Isis*, **56**, 26.

———. 1971. *The Development of Physical Theory in the Middle Ages* (Ann Arbor: University of Michigan Press).

Westfall, R. S. 1971. *Force in Newton's Physics* (New York: Neale Watson).

———. 1971. *The Construction of Modern Science: Mechanisms and Mechanics* (New York: Wiley).

———. 1982. *Never at Rest: A Biography of Issac Newton* (Cambridge, MA: Cambridge University Press).

Whiteside, D. T. 1970. "Before the Principia: The Maturing of Newton's Thoughts on Dynamical Astronomy 1664–1684." *JHA*, **1**, 5.

———. 1976. "Newton's Lunar Theory: From High Hope to Disenchantment." *Vistas in Astr.*, **19**, 317.

Williams, L. P. 1971. *The Selected Correspondence of Michael Faraday* (2 vols.) (Cambridge, UK: Cambridge University Press).

Wilson, C. 1972. "How Did Kepler Discover His First Two Laws?" *Sci. Amer.*, **226(3)**, 93.

———. 1974. "Newton and Some Philosophers on Kepler's Laws." *J. Hist. Ideas*, **35**, 231.

Woodruff, A. E. 1962. "Action at a Distance in Nineteenth Century Electrodynamics." *Isis*, **53**, 439.

Woodward, R. S. 1895. "An Historical Survey of the Science of Mechanics." *Science*, **1**, 141.

INDEX

About the Author

STEVEN N. SHORE (PhD, Toronto 1978) is a professor of physics in the Department of Physics "Enrico Fermi" of the Univesity of Pisa and is associated with Istituto Nazionale di Astrofisica (INAF) and Istituto Nazionale di Fisica Nucleare (INFN) before which he was chair of the Department of Physics and Astronomy at Indiana University South Bend. His previous books include *The Encyclopedia of Astronomy and Astrophysics* (Academic Press, 1989), *The Tapestry of Modern Astrophysics* (Wiley, 2003), and *Astrophysical Hydrodynamics: An Introduction* (Wiley/VCH, 2007). He is an associate editor of *Astronomy and Astrophysics*.